GONE TO GROUND

Intersections: Environment, Science, Technology

Sarah Elkind and Finn Arne Jørgensen, Editors

GONE TO GROUND

A History of Environment and Infrastructure

in Dar es Salaam

EMILY BROWNELL

University of Pittsburgh Press

Published by the University of Pittsburgh Press, Pittsburgh, Pa., 15260

Library of Congress Cataloging-in-Publication Data

Names: Brownell, Emily, author.
Title: Gone to ground : a history of environment and infrastructure in Dar es Salaam / Emily
Brownell.
Other titles: Intersections (Pittsburgh, Pa.)
Description: Pittsburgh, Pa. : University of Pittsburgh Press, 2020. |
   Series: Intersections: environment, science, technology | Includes
   bibliographical references and index.
Identifiers: LCCN 2019053530 | ISBN 9780822946113 (cloth) | ISBN 9780822987451 (ebook)
Subjects: LCSH: Urbanization--Tanzania. | Dar es Salaam (Tanzania)--Social
   conditions--20th century. | Dar es Salaam (Tanzania)--Environmental
   conditions--20th century. | Dar es Salaam (Tanzania)--Economic
   conditions--20th century.
Classification: LCC DT449.D3 B76 2020 | DDC 967.8232--dc23
LC record available at https://lccn.loc.gov/2019053530

Cover art: Women participating in the building of Kibamba Road, Dar es Salaam. Date unknown.
Courtesy of the Tanzanian Information Services.
Cover design: Melissa Dias-Mandoly

For my parents, Phil and Kathy

CONTENTS

ACKNOWLEDGMENTS

Much like the open-ended nature of research itself, I could have never predicted who I would meet and come to rely on as this project unfolded. In this regard, I have been so lucky. An immense number of generous people have in a variety of ways helped me articulate and shape this book long before I knew myself where it was going. This support started at the University of Texas at Austin, where I was a graduate student. The African history cohort there was a singular community. I benefited in uncountable ways from my time with Tyler Fleming, Roy Doron, Kwame Essien, Saheed Aderinto, Danielle Porter Sanchez, Charles Thomas, and Matt Heaton. We all grew as scholars and members of this community under the guidance, wisdom, and generosity of Toyin Falola, who I knew I could call on for just about anything. My gratitude extends to his family and particularly his wife, Bisi Falola. Not only did they cultivate a rich intellectual community out of their living room, they fed me and cared for me on numerous occasions. I could not have asked for a better mentor than Erika Bsumek. She has been unflaggingly supportive of my work when I needed it most. She has also, unbeknownst to her, become my model of what kind of academic I hope to be, with her diverse engagement, commitment to people, and vulnerable honesty about the scholarly life.

Austin would not have been the same without Shannon Nagy, Ross Otto, Micah Goldwater, Brencho Hughes, Sally Bergom, Grant Loomis, Lisa Gulesserian, Nick Gaylord, Chris Heaney, Hannah Carney, Kyle Shelton, Kate Vickery, Amber Abbas, Chris Dietrich, Tanvi Madan, and Sarah Steinbock-Pratt.

In Tanzania, my research had a rocky start, as is so often the case. I am grateful for the help and guidance early on of Wolfgang Scholz and Ardhi University, where I was graciously given affiliation. I am ever thankful for the staff at the National Archives of Tanzania, the East Africana Library at the University of Dar es Salaam, the Ardhi University Library, and the Tanzania Commission for Science and Technology.

I could not have made progress in conceptualizing this book without the work of my research assistant, Kassim Kindinda, who patiently helped me with my oral history interviews. Thank you to Roxanne Miller, Jamie Yang, and Tom Pyun, for being my constant dinner companions and support system during my research.

At different stages, this project benefited greatly from the financial support from the Boren Foundation, the Program in Climate Change and African Political Stability at the Robert Strauss Center at University of Texas at Austin, the Max Planck Institute for the History of Science, the University of Northern Colorado, and the University of Edinburgh.

At crucial points in the research and writing process, I am grateful for conversations with Mohamed Halfani, Kleist Sykes, David Mwamfupe, Sara Jackson, and feedback from Thaddeus Sunseri, Jim Brennan, Greg Maddox, Catherine Boone, Richard Stren, and Paul Sutter.

I am also grateful for the two anonymous reviewers who gave me feedback on my manuscript. One reviewer in particular took an immense amount of time and care to improve this book and I remain in debt to the generosity of the peer review process.

I am also thankful for the support and encouragement of my colleagues at the University of Northern Colorado, including Joan Clinefelter, Fritz Fischer, Aaron Haberman, Robbie Wies, Stephen Seegel, Mike Welsh, Jacob Melish, Jiacheng Liu, Corinne Wieben, and TJ Tomlin. I could not have asked for a more welcoming place to grow as a teacher and scholar after finishing my PhD. My gratitude to TJ in particular goes well beyond being a good colleague and carpool buddy. The Tomlins (TJ, Katrina, Carlyle, Luella, and Wendell) became my surrogate family in Fort Collins, giving me ample opportunity to not always think and worry over my book and instead listen to records, watch movies, and talk for hours over dinner and boxed wine.

This version of the book really only came together when I was a visiting scholar at the Max Planck Institute of Science (MPIWG) in Berlin in 2015–2017. In what ended up being a year and a half of residency in Berlin, I grew immensely as a member of Department III and my book took on a very different shape as a result. The privilege of time to work on my book also came with the daunting task of staring down the computer screen every day with little to justify distraction. Navigating that reality was only possible with the conversation and camaraderie of an incredible bunch of women at MPIWG: Tamar Novick, BuYun Chen, Sarah Van Beurden, Sarah Blacker, Kavita Philip, and Alina-Sandra Cucu. Dagmar Schaefer's leadership in Department III and her kindness and encouragement of my project was essential. I am especially in debt to the library staff at the institute, in particular Ellen Garske, who I am convinced is part wizard due to Interlibrary Loan abilities. This was an essential resource as my book continued to change in unexpected ways.

I also want to thank the editorial staff at Pittsburgh University Press, in particular Sandy Crooms and Amy Sherman, whose expertise and care have both helped this book immensely along the way. I am also grateful to Sarah Elkind and Finn Arne Jørgensen for inclusion of this book in the Intersection series.

I began my new position at the University of Edinburgh in the School of History, Classics and Archaeology in the fall of 2018, just as this project was coming to an end. I am thankful for my new colleagues who have helped me feel at home here, particularly Meha Priyadarshini, Emma Hunter, Jake Blanc, Martin Chick, and Emile Chabal.

Thank you to my family. My mother, Kathleen Greathouse, and my father, Philip Brownell, have always been unquestioningly supportive of where this project and career have taken me, even when it was very far away. My brother, Ben, is a model for the hard work of wrestling with ideas and aspirations that take time to coalesce. It was our family's adventurous travels at an early age that led me here.

Lastly, thank you to my partner, Kevin Donovan. When we met (on the side of the road in Dodoma, Tanzania), I was trying to jump back into what felt like a stale dissertation that had sat a bit too long on the shelf. As I began pulling it apart and putting it back together, I was wrestling with deep insecurities about my work and its worth. He read every chapter several times. And despite facing the double bind of my insecurities and defensiveness, he offered up an immense amount of intellectual and emotional labor to help me along the way.

GONE TO GROUND

Introduction

In the beginning of the nineteenth century, the town that would become Dar es Salaam was a small fishing village named Mzizima. Located just below the curve of the East African littoral where the island of Zanzibar sits near the continent's coastline, the village and region were mostly populated by Shomvi and Zaramo people. But as trade transformed Zanzibar into a major cosmopolitan hub of the Indian Ocean, Mzizima transformed into Dar es Salaam, a modest Swahili trading post.[1] Zanzibar's Sultan Majid first designed, laid out, and built up the town and its harbor in the 1860s. It subsequently became home to a polyglot population of Arab, Persian, and Swahili merchants while remaining predominantly Uzaramo.[2] In the late nineteenth century when the Germans claimed the region as their colony of German East Africa, they made the small coastal port their capital.[3] The British then claimed it as their capital in the aftermath of the First World War. Since its very beginning, Dar es Salaam has been profoundly African *and* international. And ever since Dar swallowed up the village of Mzizima, its environment has reflected a persistent tension and collaboration between the "city" and the "village."

While this interplay between city and village is a hallmark of African urbanism, it also has a particular history in Dar.[4] The lurking presence of the rural might have first developed as a product of the city's flat topography at its center which then attenuates into gentle hills that radiate into the hinterland. Reaching from the ocean into the valleys of these hills, a system of creeks spread across the city's landscape, waxing and waning in the rainy and dry seasons and changing the city's contours—sometimes dramatically— as a result. With these transient flows, the city's expansion has never been straightforward but rather punctuated by watery boundaries that disrupt any urban coherence.

The tension between city and village, though, is also a social artefact of a racist colonial history that frequently relegated Africans to the town's periphery, beyond its planned center. When first the Germans and later the British made it their colonial capital, both administrations entrenched racially segregated neighborhoods into urban planning law, frequently utilizing these creeks and hills as dividing cordons sanitaires.[5] Similar environmental boundaries mark the history of many colonial cities across the continent where authorities often mandated that Africans remain in peri-urban "villages" or unplanned areas off the official urban map. These measures were frequently entrenched in a rhetoric of public health, but administrators also aimed to avoid the costs of providing durable urban housing. Colonial officials worried, too, that "natives" who relocated permanently in cities would lose their moral bearings if unmoored from "tribe" and "village."[6] For the modest population of Africans who did find access to planned urban neighborhoods by the late colonial period, they had to justify their presence with proof of employment or potentially face expulsion from the city.

But even after independence, when Tanzania's first president, Julius Nyerere, decried the segregated legacy of the city, the ideological specter of the village still haunted urban development. No fan of Dar, Nyerere held up "the village" as a moral virtue, urging and sometimes forcing residents of Tanzania's biggest city to leave for surrounding villages in order to enact a socialist rural future. Echoing policies of the late colonial period, regional government officials in the 1970s and early 1980s rounded up unemployed residents and dropped them off in Dar's periphery to become part of newly forming villages. Nyerere's colorful, blunt antiurban rhetoric plays a central role in many histories written about the postcolonial city. That Dar developed such a dynamic cultural life while state development policies valorized rural settlement offers a rich tension to these scholarly works.[7]

However, if we turn our attention to the city's built environment rather than its political and cultural life, the supposed antipodes of city and country dissolve and begin to seem more colluding than contrarian. These landscapes are less distinguishable from one another by this period than perhaps ever before.[8] Despite generations of authorities drawing moral boundaries between the village and the city, residents of the region routinely traversed both worlds for their own purposes and projects. In the late 1950s, when the retired colonial officer J. A. K. Leslie was tasked with writing a book on the welfare of the city's African population, he remarked, in passing, that locals exploited the opportunities of living somewhere between the village and the city. More specifically, Leslie noted that many Zaramo were leaving their villages in the coastal region in periods of drought and seeking economic opportunity in the city. But rather than an inexorable migration toward the center—or what scholarship on cities has tended to make synonymous with "urbanization"—Leslie observed that some Zaramo resisted or were not willing to "go completely urban by taking a job and relying on it exclusively for income."[9] Instead, he noted, many would "hang on by 'going to ground'" in the city's rural periphery. A British colloquialism, to "go to ground" means to lie low or hide out from authorities, and in fact the Zaramo were frequently seeking refuge from officials during periods of strikes, unemployment, or routine purges from the city. But beyond serving as a place of refuge, the periphery was also a place of temporary material relief, just as the city was at other times. Indeed, while many scholars of Leslie's generation saw this back-and-forth as a metaphor for Africans' alleged failure to modernize, it was precisely what becoming "urban" looked like in Dar es Salaam: an ongoing process of negotiating the opportunities and struggles of the city through seeking the relief of rural resources rather than a finite transition from the village.[10]

Struck by the evocative image of "going to ground," this book takes up Leslie's brief aside and reworks it as a conceptual approach to writing the environmental history of Dar's changing urban landscape in the 1970s and 1980s. I also use it to write against the popular and scholarly penchant to separate histories of the rural and the urban. In blurring these landscapes and categories, I seek to be part of a long tradition of pushing back against this distinction in environmental history.[11] I argue that it was during a period beginning roughly in 1973 that a new era in Tanzania's urban history began, as the city rapidly grew while also being gripped by an unfolding economic crisis and fracturing of urban infrastructures. In response to these circumstances, Dar's citizens coped much like the Zaramo in the 1950s; families increasingly made their lives in transit between the city and its periphery, sometimes also to evade the state. In doing so, they were quite literally turning to the ground to make life possible when they were either short on cash or other urban shortages broadly persisted. They exploited the coastal region's natural resources to shape their lives in Dar, relying on the city's outskirts to plant small *mashamba* (farms) or to seek out building materials for their houses, goods to sell at markets, or charcoal for cooking the evening meal.

In revealing how families and neighborhoods stitched together city and country, this book considers how Dar es Salaam's environments and infrastructures reflect the accumulation of everyday acts of provisioning for urban lives. It was through family labor as much as corporate or state labor that both the hinterland and the city were transformed. Environmental histories of cities tend to examine longer periods of time and narrate scaled, capitalist transformations of "nature" into urban infrastructures and space—the myth of the modern city is that it becomes a place where communities are insulated and ultimately alienated from direct relationships with nature. In framing the heart of this book around little more than a ten-year period and considering everyday struggles at a time of economic crisis, what instead comes into focus is how urban communities became not more cut off from "nature" but rather more entrenched in it, defying easy categorization.

But traversing between city and periphery is not the sole focus of this book. I also conceive of "going to ground" to include how urbanites made their own plans and infrastructures for life in the city when municipal services, factories, public transportation, and institutions of planning were routinely, exasperatingly out of commission or grounded. In this period, urban infrastructures suffered from material shortages, lack of expertise, and infrequent maintenance, causing many to be frequently suspended. Facing these periods of disrepair, urbanites developed a diverse repertoire of ways to inhabit the city while also venting their frustrations publicly about the state of Dar's physical infrastructures. By tweaking ailing infrastructures and technologies or simply navigating around disrepair, urban communities reshaped their built environment in profound ways.

The state also struggled to provision for its citizens. Urban authorities similarly faced the necessity of developing new strategies for dealing with infrastructural failure and determining who was to blame. Likewise, when several of Dar's factories went offline during the 1980s due to dwindling foreign exchange funds, parastatal factory managers had to improvise new forms of production. These improvisations turned to local raw materials to avoid expensive imports and became new ways to conceive of "nation building," quite literally from the ground up. These solutions were practical and ideological, shaped by Tanzania's membership in the Third World and the state's desire to create a new global economic order, highlighting once again the city's position mediating between the local and the international.[12] Thus I also engage with one final definition of "going to ground" in this book: the emergent hope of a new, grounded path to development that accompanied the heady insecurity of life in Dar. This was the hope of creating a rooted, local, and self-sufficient city and nation from the intellectual and political collaborations of the Third World. In all these manifestations of "going to ground," Dar's residents and the state came to rely on the city's environment in new ways and in turn they shaped the city profoundly.

Urban Growth and Writing Histories of "Third World" Cities

Many of the activities that constitute "going to ground" trouble the trajectory of prevailing narratives of urbanization. In the existing historiography, urbanization remains a process in which cities become more "citylike" over time. Predominantly through histories of the industrializing West, we have come to a loose consensus of what materially and environmentally makes up a city and demarcates it from its enduring scholarly foil, the country. African cities have long been sidelined in this history, in part because they have been measured against these notions of "cityness" that they do not reflect, particularly in their frequent construction out of "crude brick, straw, recycled plastic, cement blocks, and scrap wood."[13] Their material presence has left them seemingly in a perpetual state of becoming. As a result, scholars have theorized with "First World" cities while viewing "Third World" cities like Dar through the "lens of developmentalism."[14] This literature tends to focus on "capacities of governance, service provision, and productivity" with an implicit understanding that these cities need to be fixed.[15] To explore how communities conjured their lives in Dar at a time when both materials and political will were in profoundly short supply is then also a call to think more broadly about what and who constitutes urban environments. How might the routines of families moving between the city and its periphery actually redefine what counts as the space of a city? And as the "expectations of modernity" in Tanzania fall short of the state's original imaginings, how does the state also participate in rethinking what materials and infrastructures constitute a "modern" city?[16]

While there is an immense history still to be written on the environments of postcolonial cities, historians have nevertheless still been telling stories about these landscapes.[17] In the instances when places like Mumbai, Lagos, Dhaka, Dar es Salaam, and Cairo show up in environmental histories, they tend to get combined into peripatetic accounts of "global cities" rather than serving as the subjects of separate monographs.[18] In these accounts, such cities can stand in for the problems of the developing world writ large. This is particularly true with the emergence of "megacities" in the twenty-first century that have come to symbolize an archetypal environment of the Anthropocene. In this literature, the so-called Third World cities signal the perversions of urbanism and planetary well-being, even while aspects of them are celebrated.[19]

These confounding urban landscapes are frequently first introduced to readers through numbers. In the very term *megacities*, we are first drawn to the dramatic and rapid expansion of urban life in the Global South. Scholarly accounts of these cities begin by reciting the statistics of their unparalleled growth in the last fifty years. In this formulation, the numbers would tell us that Dar is no exception: between 1968 and 1982, a timeframe that this book sits roughly within, the surface area of the city multiplied by five.[20] This statistic, and an annual growth rate of 7.8 percent during this same period, were facts that I would routinely tell people when asked why I was writing an environmental history of the city.[21] After reciting these numbers enough, however, I became quite critical of my own recourse to growth as justification for Dar's importance. Growth—the more dramatic the better—struck me as the predominant mode by which environmental historians and urban scholars justify their engagement with cities of the Third World, and this justification is prone, problematically, to confirming rather than challenging readers' ideas about the Global South. In lieu of a richer introduction, a city's "sprawling," "teeming," and "unremitting" expansion frequently exists on the page before the city itself does: lost in scale, there is no "there" there. The sordid extremes of Third World urbanism should not become its scholarly contribution.

I am certainly not the first to critique these approaches to urbanism in the Global South. In the last fifteen years an emerging scholarship has begun to rewrite the place of African cities in urban studies.[22] Anthropologists and geographers in particular have been at the forefront of this new work.[23] There is a lively sense of African cities as iconoclastic and hard to pin down captured even in the titles of these new theoretically oriented works.[24] Additionally, a collection of books by cultural and social historians have also enriched our understanding about sexuality, gender, pop culture, and race in the African city.[25] A history of Africa's built environment, however, is not foregrounded in this recent burst of historical work.[26]

With more than half the world living in urban areas, writing more expansive environmental histories of the Global South is vitally important. These histories could ground our understandings of a period "often categorized as unprecedented and therefore somewhat evasive of historicization."[27] Without highlighting other stories along with narratives of growth, we are left with an understanding of these cities as cautionary tales of overpopulation and underdevelopment. The very notion that we now live in the moment of the "great acceleration" can conjure a sense that these urban landscapes are the result of an unfolding algorithm rather than revealing of their histories and environments.[28] Clearly, just like in nineteenth-century London or Paris, rapid urban expansion today has led to the accompanying problems of pollution, waste, sanitation, and sprawl. While exploring some of these issues, this book suggests that there are other environmental stories to be told about how people build homes, provision their lives in the city, and connect the challenges of the urban environment to both personal and national aspirations.

In seeking to enrich our understanding of these urban environments and infrastructures, this book walks a fine line. On the one hand, the following thematic chapters represent my attempt to banish the "specter of comparison" that has haunted Third World cities and instead "world" Dar's changing landscape at a time of deep anxiety about global environments.[29] On the other hand, these same chapters engage with the narratives of crisis that have come to exclude African cities from historical narratives other than as sites of economic, demographic, and environmental catastrophe.[30] The 1970s were rife with the pronouncement of crises both globally and particularly in East Africa: the oil crisis, the wood fuel crisis, and the urban crisis lurk in these chapters. By the 1980s Dar es Salaam's landscape was shaped most fundamentally by what residents remember as crippling economic austerity that left them planting their own food, disposing of their own waste, improvising transportation, fueling their own households, and building on unzoned plots of land. By placing these unfolding events as central to each chapter, this book is an environmental history of an economic crisis as well as a city. Indeed, many urbanites might most readily recall how they mitigated against crisis through their engagements with Dar's swiftly changing urban ecologies and infrastructures.

As a result, *crisis* is not a word nor a sentiment that I can avoid but it also cannot go unexamined, as I will return to in the conclusion. Like narratives of uncontrolled growth, by engaging with crisis narratives I risk reinforcing a problematic and popular view of the African continent. This is distinctly not my goal. Rather, what I intend here is two-pronged. First, I argue that the quotidian types of interruption that the following chapters focus on might be augmented in Dar, but they are part of everyday life in all cities: we know this in the frustrating commute to work when the subway goes offline, when fuel costs spike, or when sewage lines back up and disrupt our daily routines.[31] Recent works on infrastructure by science, technology, and society (STS) scholars offer an important reevaluation of how we tell the history of cities. While environmental historians have narrated the construction of massive infrastructures such as dams, highways, electrical grids, and sewerage systems, STS scholars who draw attention to how these infrastructures are subsequently used, repaired, and reinvented offer a crucial second half of the story.[32] Postcolonial cities, with their skeletal budgets, had fewer official backup plans. When the bus broke down or the electricity went off, urban residents, workers, and state officials were more routinely forced to improvise. With enough repetition, these improvisations shaped urban landscapes in ways that are still unfolding. In this way, the crisis of municipal services that urban residents faced by the 1980s might be more dramatic than those faced in other cities, but they inform the history of all cities as ongoing places of repair and reinvention.

Second, to avoid the constantly lurking language of crisis that I encountered in my research would leave a key topic unexamined. How did the discursive construction of these crises shape how urban residents navigated the city? How did the perpetual threat of food or fuel shortages shape Dar's landscapes? As I draw out in my final chapter on the wood fuel crisis, sometimes a crisis foretold never actually arrived. And yet the pretense of disaster nevertheless shaped and facilitated international intervention in the lives of local communities. Long after a "crisis" is over, its aftereffects also continue to shape how outsiders see cities like Dar and who is blamed when problems arise.

African cities are not landscapes that have only emerged out of the failure of "proper" forms of urban life to take hold, or out of "informal" rather than "formal" urbanization. And yet, residents of these cities have historically been forced to deploy creative responses to the foreclosure of plans, infrastructures, and imagined futures. In examining such moments of recalibration, scholars should resist simply valorizing Africans as "resilient" subjects who can overcome all obstacles. Nevertheless, the ways in which urbanites dealt with the difficulties of everyday life gives shape to much of this book.[33] As Gabrielle Hecht has warned historians of technology in Africa, "it's important not to be seduced by the romance of creativity. We mustn't overlook conditions of scarcity. Those conditions matter." To the historical actors we study who were navigating scarcity, "inequality matters," too: "It's not that they prefer this state of affairs. It's that they're making do with what they have at hand. That's a delicate interpretive balance, which both Africanists and STS scholars have to walk when they're traveling down this path in conversation with each other."[34] I have tried to walk this path carefully in the following chapters.

Environmental History with African Sources

Because postcolonial cities have different histories than places like Chicago or London, they also leave behind a different palimpsest of sources. The following chapters thus take their shape from the sources and methodologies of African history as much as environmental history. The historian Luise White recently urged historians of postcolonial Africa who tend to gnash their teeth over the gaps in sources to instead take the "mess" of postcolonial archives as their "starting point." Rather than trying to madly patch over the "hodge-podge" nature of the historical record, White argues, "the gaps and the fissures are not simply problems or absences in the archival record" but help us understand "states and policies and plans as a bricolage."[35] This resonated with me and influenced what I have written in this book. It also resonates with how the Congolese author Sony Labou Tansi describes the continent's cities. Invoking Tansi, AbdouMaliq Simone notes that urban Africa reflects the "African love affair with the 'hodgepodge'—the tugs and pulls of life in all directions from which provisional orders are hastily assembled and demolished, which in turn attempt to 'borrow' all that is in sight."[36] The following chapters wrestle with, and hopefully capture, both of these patchworks: the postcolonial archive and the postcolonial city.

In piecing together the story of the 1970s and 1980s, I was only able to uncover a very modest municipal archive.[37] The history of Dar es Salaam is not the history of an unplanned city, but it is one that must be written from beyond the planning archives. As I lay out in the first chapter, Tanzania's president, Julius Nyerere, and the state's municipal planning apparatus actively turned away from Dar es Salaam in the 1970s, leaving it to be developed predominantly by its residential communities. To confront the paucity of municipal records, I first conducted oral history interviews with residents of the peri-urban community of Mbagala. Mbagala is one of many neighborhoods that emerged in the 1970s as urban residents looked for cheap land and began moving to the outskirts of the city.[38] My interviews with these men and women helped me understand how residents settled and unsettled repeatedly in the city, sometimes within and sometimes outside of official channels. These interviews disrupted the notion of urbanization as a one-way process and highlighted how urban residents shaped Dar's surrounding environment in the process of making their lives possible in the city. But rather than proving that Dar was a quintessentially "unplanned" city, these interviews showed me how central the state remained in orchestrating an ethos of "self-help urbanism." Despite the importance of these interviews in my own research process, they did not ultimately constitute a major source in most of the following chapters. I conducted them early on in my research and as a result, they informed me far more than they might show up in the following pages, where they are mostly used as illustrative.

Perhaps the most prolific source in this book is Dar's rich newspaper archives.[39] These newspapers became a crucial way to flesh out stories about urban infrastructures and the communities that shaped them. Tanzania's newspapers routinely chronicled when production at the cement factory stalled, where water pipes burst, or the ongoing frustrations of intermittent bus service. In some instances, newspapers also functioned as a prescriptive space, publishing how-to articles for navigating shortages or reappropriating overlooked materials. By publishing reader letters, the *Daily News* in particular provides a window into how urbanites reacted to their changing city. Those who wrote to newspapers were "performing a certain kind of public self" that revealed both the impatience and aspirations of urbanites regarding how their city should function.[40] There is an immediacy as well as poignant mundanity to these accounts that cannot be recaptured years later in oral histories.

The press also exerted its own pressures on the state to address certain urban problems and thus must be understood as an actor in their own right too, shaping a discourse about a "dirty" city. In one particular incident, editors from the *Tanganyikan Standard*, frustrated with an expanding pothole near their offices in the city center, decided to print comical photographs of the offending spot. First they printed a photograph of children playing in the pothole after it had filled up with fetid water. A few days later, they snapped a photo of small boats floating on the expanding pond and published it, appealing to the city to fix the pothole. Finally, and most absurdly, they staged a fisherman "complete with goggling gear and spear gun" holding up a large fish in one hand."[41] Shortly thereafter the city began repairing the road. These letters and photographs documenting the city give voice to the frequently banal environmental forces of water, mud, sand, and salty sea air that lay at the heart of debates between citizens and the state about the condition of the city.[42]

Finally, I have also assembled for this book a transnational archive of technocratic gray literature of development.[43] This literature includes in-

structional manuals on how to make burnt bricks; scientific studies on deforestation and charcoal use; urban master plans; project reports from the
World Bank, International Monetary Fund, and United States Agency for International Development; transnational expert training programs; proposals
for Third World technological transfer; and a wealth of studies run by students and faculty at the University of Dar es Salaam. These sources shape the
story of Dar's built environment as fundamentally transnational, even when
grounded in a relationship between the city and its surrounding resources.
These documents also capture the 1970s as a moment when Tanzania, along
with many decolonizing nations, were forced to recalibrate their vision of the
future as conditions changed. Within the pages of these studies, this urgent
need to rethink assumptions of modernity are captured in both practical and
ideological terms. These attempts to reconsider the future were built on transnational economic, intellectual, and material connections. The Tanzanian
state framed its own pursuit of economic and political sovereignty as part of
a larger African and Third World struggle for decolonization and independence. Using this literature to look outward from Dar connects the dilemmas
facing the city to the larger hopes and fears of the postcolonial world. When
read together, these individual studies and pamphlets chart a trajectory from
large-scale infrastructure projects and plans shaped by Western expertise
to more modest plans and "appropriate technologies" that turned instead
to other Third World countries, particularly India, for assistance. This shift
evokes an evolving critique of the West as peddling an unsustainable model
of development particularly as oil prices and commodity prices rose again
in the early 1980s. The Tanzanian state continued to reassess their own use
of environmental resources even as the conditionalities of lending agencies
bore down on them. They sought a future grounded in their own resources as
this became less and less possible due to crippling debt.

Outline of the Book

In taking shape around a hodgepodge archive, this book is organized themat-
ically to represent not institutions or epochs but the struggles and opportunities of the city in a way that might resonate with those who called Dar home
during this period. By focusing on quotidian processes of city making, I have
aimed to keep my own sensibilities grounded, not retreating too frequently
to the bird's-eye view of a city that most residents inhabited by walking. The
opening chapter, however, is an exception. I begin by locating Dar geographically and politically within the new nation and within President Nyerere's
vision of Tanzanian socialist development known as *ujamaa*. I also place Dar
temporally within a moment when cities globally are becoming seen as sites of
crisis rather than paths to modernity. In this context Dar lost its status in the
early 1970s as Tanzania's capital, to be replaced with the new planned capital

in Dodoma, in the heart of Tanzania. Considering this loss of official status as well as the plans for the new capital fleshes out how the state envisioned the future of its most populous city. And yet while the president in particular hoped to diminish Dar's symbolic and economic importance, Tanzanians nevertheless still moved there in near record numbers. These tensions—and ultimately contrasting aims—of the state and its citizens highlight not just the personal preference of families, but the material constraints placed on decolonization by the legacy of uneven development.

In chapter 2 I turn to how arriving families found land and material "belonging" in the city. As part of a much longer colonial history of precarity for rural migrants, finding a foothold in the city in the 1970s became increasingly about looking outward to the periphery. This was due to failures in housing provisioning as well as pervasive commodity shortages that prompted families to grow their own food. It was also due to state efforts to remove all "unemployed" people from the city to the surrounding countryside. But claiming space in the periphery also became the strategy for families who had made Dar home for much longer. As the city's outer edges emerged as a new center of activity, Dar's urban core also became a heterodox, ruralizing environment.

Chapter 3 looks at the politics and practices of building in Dar, focusing on the materials that constituted the average improved "Swahili-style" home in the city: mud bricks and concrete. At the center of this chapter is the story of the state's decision to build a cement plant outside of Dar to serve expanding construction needs and to signify the nation's arrival in a modern future. But as production at the plant faltered due to a variety of factors, the state was forced to consider the merits of alternative building materials and methods, resignifying mud bricks as part of an alternative Third World modernity and rejecting concrete as imperial. Regardless of state rhetoric, though, these materials still had to be taken up and made real by family builders. As a result, the materiality of Dar's houses reveals both state and family aspirational narratives about the future and how they were mitigated by the realities of the present.

Chapter 4 turns from materials to temporalities of the city. As many reading this book will know, traversing African cities can be punctuated by long bouts of waiting. But waiting shapes urban livelihoods in ways far beyond queueing for the bus or stalling in traffic. This chapter pairs a brief history of roads and transit in Dar es Salaam with an exploration of what is work (and what is loitering) in the city. How did infrastructures of transportation—expanded in the postcolonial period with the socialist urban worker explicitly in mind—shape definitions of labor in the city and shape urban landscapes in ways that unfolded daily, seasonally, and ultimately over decades? Waiting for the bus, waiting for a job, and waiting for "development" were all affective states that shaped the conversations, ideologies, and environments that constituted urban life.

Chapter 5 looks at how the city's material flows of waste, food, and manufactured goods were dramatically reconfigured following a drought in 1974 and subsequent economic struggles in the early 1980s. As urban residents navigated an unfolding crisis, they had to find new channels for food staples. Urban authorities meanwhile struggled with sanitation services as trucks went unrepaired and petrol prices spiked. Cascading shortages of foreign exchange also forced factories to rethink their raw materials. These struggles sparked larger conversations about production and consumption under scarcity: what should Third World manufacturing look like, and where was the sometimes imperceptible line between citizens who conserved and citizens who hoarded?

The final chapter continues to consider the question of what a decolonized nature and economy would look like in Tanzania and how Dar's urban crisis provoked new practical and ideological visions of the future. This chapter chronicles the city's charcoal market within the broader global moment of the fuelwood crisis that emerged alongside the oil crisis. As environmentalists in the Global North began to worry about imminent deforestation from peasants in the "Third World" using trees for fuel, both the Tanzanian state and its urban citizens turned to charcoal as an alternative, autarkic fuel source. Mirroring the first chapter's turn to global planning narratives, this final chapter also pans out to place Dar within a global moment, this time the emerging environmental movement.

These chapters are not always orthodox or explicit in their focus on how nature shaped the city or how those in the city shaped nature. In reflecting the urgency of certain narratives that emerge in my sources, they are instead about how the city's built environment emerges as the result of other concerns of the postcolonial period. Due to other ideological and financial priorities, improving urban infrastructures was never foremost on the state's agenda. Urbanites thus had to give their own shape to Dar's environment while hoping to secure access to the relative advantage of the city. Walkers and hawkers shaped the city's streets and sidewalks in the absence of functioning public transportation; factory workers shaped its valleys and riverways by growing food on their weekends; university professors cut into its forests to sell charcoal and subsidize their salaries. In telling these stories, I hope this book pushes some of the boundaries of urban environmental history in service of bringing new cities and narratives into the fold.

Chapter 1

DECENTERING DAR

On the eve of the 1970s, Dar es Salaam was caught up in a vertiginous global moment. The coastal city halfway up a continent in the throes of liberation had become a vital hub of Pan-African resistance and intellectual life, particularly for its neighbors to the south still under white minority rule. At the same time, the seemingly magnetic pull to the city for rural Tanzanians was becoming a thorn in the side of the newly socialist state. While Dar had trebled in size in little more than a decade, the state was doubling down on its vision of a rural, agricultural future.[1]

For African cities more generally, the pendulum of national and international esteem seemed to be reversing course. From manifesting hopes of modernity and development, cities like Dar now seemed to embody fears of chronic poverty and demographic disaster. Rapid urbanization had become "perhaps the most dramatic social phenomenon that marked the end of the colonial period in Africa."[2] Its proliferating effects shaped a generation of scholars who began puzzling over the alarming rates of rural-urban migration, the problem of "urban bias," and the expansion of "slums" at the expense of agricultural production.[3] While these anxieties were rooted in changing African realities, they also reflected a global skepticism of urban life that had emerged by the late 1960s. As Third World cities swiftly grew, their counterparts in Europe and the United States fell into disrepair and disrepute. Perhaps most memorably captured in post–studio era Hollywood cinema, these cities evoked social decay, infrastructural neglect, pollution, and poverty.[4] Paris, New York, Detroit, and London were sites of what became referred to as an "urban crisis," marked by race riots, radical social movements, and white flight. Urban planners were also turning away from the spectacle of the skyscraper and seeking out a smaller scale. Eschewing the crowded tableau of cities, these planners embraced the space of the planned-from-scratch New Town, imagined to be self-sufficient and in harmony with its natural surroundings.[5]

Alongside this growing global anxiety over the future of cities emerged the environmental movement. The movement gathered force by the early 1970s around fears of unsustainable planetary growth and declining food production. Sparked in part by the 1973 Arab oil embargo and the newly popularized field of ecology, a vanguard of "sober prophet[s] of impending doom" made their way on to late-night talk shows and penned newspaper editorials expressing neo-Malthusian concerns that planet Earth was on the verge of collapse.[6] Perhaps best captured in the work of Paul Ehrlich, many of these anxieties about the planetary future took specific aim at the decolonizing world, portraying cities in particular as the new demographic ground zero.[7] In his bestselling book, *The Population Bomb*, Ehrlich's first chapter, "The Problem," opens with a description of a taxi ride through Delhi, where "the streets seemed alive with people. People eating, people watching, people sleeping. People visiting, arguing, and screaming. People thrusting their hands through the taxi window, begging. People defecating and urinating. People clinging to buses. People herding animals. People, people, people, people."[8] No doubt these images and anxieties of crowded cities helped foster a new emerging development ethos that sat at the other end of the scale: "small is beautiful."[9] If the environment of these cities was not yet of much global concern, the notion that these cities were themselves global environmental threats quickly gained traction. Development experts and international financial institutions such as the World Bank and the International Monetary Fund shifted their priorities from funding massive infrastructure projects to focusing their aid on agriculture. The prototypical small-scale farmer emerged as the most deserving recipient of aid, reaffirming the importance of the rural and the potential ecological disaster of unchecked urban migration.

The new nation of Tanzania emerged as a darling of the development world in this anxious global moment. The Tanzanian state's vision of the nation's future seemed to reflect these shifting values and emerging anxieties, leading to what one African historian at the time diagnosed as "Tanzaphilia."[10] "Tanzaphilia" was the "romantic spell" cast on Western intellectuals and development experts obsessed with the potential of Tanzania's future to serve as a global example for the Third World. The colony of Tanganyika had gained its independence in 1961 when Britain handed over power to the charismatic and popular nationalist leader Julius Nyerere and his party, the Tanganyikan African National Union (TANU).[11] In 1964 Tanganyika joined with the island of Zanzibar to form the new nation of Tanzania. Six years after independence, in 1967, President Nyerere introduced one of the most striking political documents and projects of African independence: the Arusha Declaration.[12] The Arusha Declaration ushered in the new political project known as *ujamaa*, an intellectual and political philosophy of development that aimed to create a distinctly African form of socialism.[13]

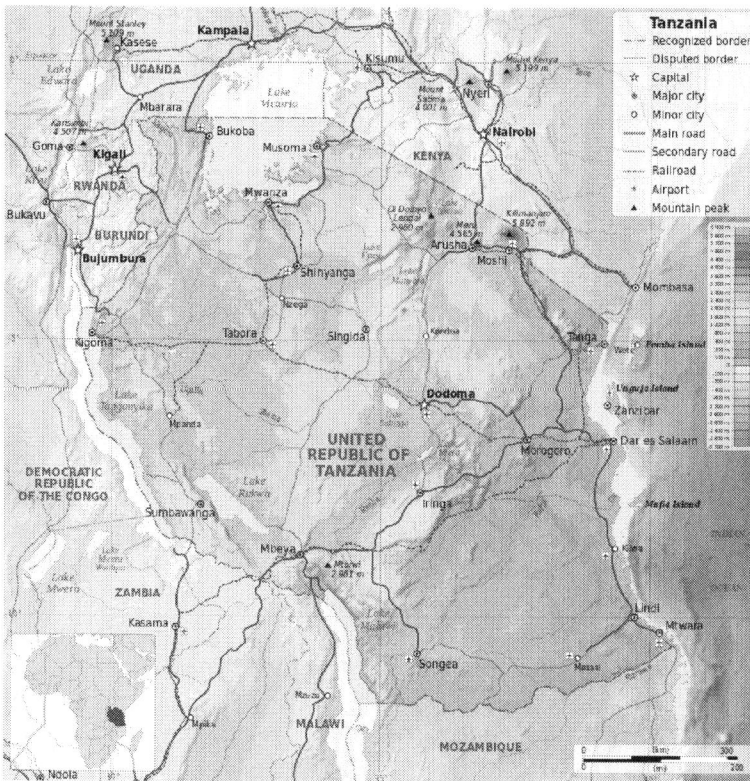

1.1. Map of Tanzania in English. Equirectangular projection. 1° N, 28° W, 42° E, -13° S. July 8, 2014. Created by Sémhur; translated by Jen. Wikimedia Commons, https://commons.wikimedia.org/wiki/File:Tanzania_map-en.svg.

Nyerere's vision of African socialism presumed and planned for a vast majority of Tanzanians to continue living rurally. Like all new African nations, Tanzania's nascent economy reflected its recent colonial status. Built around the export of agricultural goods and raw materials, ujamaa aimed to transform the colonial economy of extraction into communal production on cooperative village farms. Organizing development around rural production was both practical and ideological. Nyerere invoked Tanzania's precolonial past to advocate for a future shaped around the village and cooperative labor, but he also insisted that ujamaa was not a retreat into the African past. Rather, it offered a path beyond "the way we lived years ago."[14] As he once impatiently exhorted, "our tools are as old as Muhammad; we live in houses from the time of Moses! . . . While the Americans and Russians are going to the moon, we Africans are dancing. . . . Our friends are using their brains while ours sleep and grow fungus! They are sending rockets into outer space while we are eating wild roots!"[15] The president saw ujamaa as a way to banish the "backwardness" of Tanzania's colonial past; first the Germans and then the British had left most Tanganyikans in small settlements cut off from markets, health care, education, and transportation, while they lived in the comfort of the colonial capital. The hope of newly planned villages was in their capacity to serve as fundamental tools for modernization. Indeed, the village was not a retreat from modernity but a place where citizens "would be more accessible to experts."[16]

As outlined in the 1968 villagization scheme known as Ujamaa Vijijini, this transformation would happen through the establishment of new planned villages large enough to merit their own schools and clinics and to sustain communal agriculture.[17] These villages would connect Tanzanians to the world and vice versa. Through remaking the demographic map of the country, ujamaa would support an independent future for Tanzania, providing through agriculture the economic means for developing import substitution industries while also preserving an "African" way of life. While sometimes this transformation happened simply through the bureaucratic rechristening of an existing village as an ujamaa village, the scheme also imagined that Tanzanians would voluntarily uproot their lives and relocate in order to create new villages and thus a new nation. When this did not happen on a large enough scale, what had been voluntary became compulsory. Beginning in 1973 Nyerere ordered Tanzanians to aggregate in villages no smaller than 250 people.[18] What followed over the course of the 1970s became one of the largest relocations of Africans on the continent. In total, between eight and nine million Tanzanians were relocated to ujamaa villages, sometimes forcefully, through the destruction of their existing homes and communities.[19]

As other historians have pointed out, it is hard to find any state articulation of what urban ujamaa would look like as villagization began.[20] Beyond serving as sites of manufacturing and professionalization, cities are glaringly absent from ujamaa and the imagined future of the new nation. Instead, state policy documents and speeches reflect a nearly ever-present skepticism of urban life. This captures perhaps the central struggle that has animated state narratives of Tanzanian history and defined the character of ujamaa: the struggle between exploiters and producers.[21] In imagining a nation of communal rural production, the absence of cities from ujamaa reflected Nyerere's belief that especially Dar was the refuge of exploiters; the draw of an easier life in the city was literally draining the country of its young labor force. Beginning in the colonial period, the moral economy of senior men across East Africa also tended to consider urban areas to be dangerous places that challenged their authority and power: cities were where women became prostitutes and young men upended chiefly hierarchies through their access to labor and wages.[22] Indeed, an archetypal antagonist in Tanzanian nationalist political thought is the character of the *unyonyaji*, a vampiric stranger particularly associated with cities capable of both literally sucking the blood and fluids from Africans as well as figuratively exploiting them.[23] Even after independence, state propaganda trafficked in easy caricatures of the rich urban property owner who collected rents rather than working for a living or the poor rural migrant who eked out an existence through buying and re-selling goods, adding nothing to the export economy. Both were pilloried for relying on the productive rural farmer to keep the nation afloat in foreign currency and food.

In this chapter I want to consider how a circulating global unease about cities, particularly as it was refracted among an international community of urban planners, mixed with the Tanzanian state's own desires to focus on rural development and dislodge Dar es Salaam as the administrative center of the nation. Global fears of an unfolding "urban crisis" resonated with Nyerere's development ethos and the older regional moral skepticism of cities that has roots in the colonial era. In bringing together this global planning shift with the state's antiurban sentiment, we can consider Dar within a larger moment of pessimism about cities while also focusing on two concrete policies that reflected the state's efforts to marginalize Dar in the future of Tanzania. Also, by briefly considering international and national cultures of planning at this historical moment, I hope to explain why urban institutions of planning *do not* figure largely in a book about urban infrastructures and the built environment. Despite being perennially and derisively described as "unplanned," Dar is a city with a rich planning history and these institutions certainly did exist in the 1970s. They were, however, deeply disempowered and lacked sufficient expertise as well as the will of the state. This reality reflects the legacies of underdevelopment and skeletal colonial urban planning that shaped the futures of many African cities. But it also reflects particular state policies designed to turn state funds away from Dar in order to focus on development elsewhere at the beginning of the 1970s. This shift is reflected in TANU's decision to relocate the new nation's capital from Dar es Salaam to Dodoma in 1973, and is also reflected in the near-simultaneous decision to adopt a new policy of bureaucratic decentralization. These policies did not seal the city's fate in any decisive way, but they did dramatically shape the conditions of possibility for Dar's built environment and render many forms of urban governance relatively powerless in the face of rapid urban growth. I end this chapter by considering why, despite growing pressures to settle elsewhere, rural migrants continued to seek out the city. In the face of immense political and intellectual forces aiming to decenter the old capital, the legacy of uneven development that had created Dar meant that for most Tanzanians, the city retained its magnetic pull in times of shortage and need.

3. Dodoma, Central Location

1.2. Image showing the centrality of Dodoma. Originally from Dodoma master plan by Project Planning Associates Ltd, 1976. Source: Sophie van Ginneken, "The Burden of Being Planned: How African Cities Can Learn from Experiments of the Past: New Town Dodoma, Tanzania, *International New Town Institute*, http://www.newtowninstitute.org/spip.php?article1050.

Relocating the Capital

Readers picking up a newspaper in 1972 would have found themselves quickly embroiled in an ongoing debate about whether to relocate the bureaucratic capital of Tanzania from Dar es Salaam to Dodoma. Sitting on the edge of the coast and bathed in the humidity of the Indian Ocean, Dar was five hundred kilometers nearly due east of Dodoma and yet a world apart from its new rival. Dodoma is nestled in a very dry and rocky region marked by a persistent history of famine.[24] As a railroad hub, even as far back as 1915, a German colonial officer had suggested that the dusty town might serve as a prospective colonial capital. While Dodoma was less malarial and far more arid than Dar, the coastal city's strategic location nevertheless won out, leaving Dodoma small and relatively unimportant despite its geographic centrality.[25] But it was perhaps Dodoma's continued humble appearance that now became an argument in its favor when surveying new prospective capitals. Lying at the center of the new nation, this was a chance for regional and economic redemption for the windswept whistlestop.

As TANU regional committees debated plans for relocation, the *Daily News* printed dozens of letters from citizens across Tanzania weighing in on the decision. Many wrote in worrying about the price of such aspirations and pointing out that it seemed like an extravagant expense for a country shaping its future around the humble village.[26] And while Dar's relative overdevelopment and overprivileged status served as a state rationale for relocation, a narrative of the imminent decline of Dar also served as justification. It was time to start anew. Readers in favor of moving the capital wrote in to bemoan Dar's dilapidated state of infrastructure as a reason to go elsewhere. These letters argued that the very concept of planning—inherently future-oriented—condemned the former capital as something that it was *too late* to save. Planning had already failed and the damage was already done. "This so-called capital does not justify the status 'city,'" one Dar resident wrote in. "The whole town is an eyesore and lacks proper planning. The town's architecture is out-moded. So Dar es Salaam does not qualify to a seat of government. We need a clean and nice city with modern architecture, modern roads and streets, modern recreation facilities, modern parks, modern stadiums, and all good things that go with a city. The roads and streets of Dar es Salaam are abominably narrow, dirty and do not allow for meaningful expansion."[27] Furthermore, "a number of supporters of relocation advanced the argument that the country's leaders required reinvigoration by Dodoma's ascetic setting. Its isolation would have the effect of weeding out those unprepared to serve national political goals. People in Dar es Salaam[,] it was said, 'need their roots torn up' as they have become excessively individualistic."[28]

Dar's colonial past as a racially segregated city with a deeply uneven distribution of infrastructure had left large sections of the town to expand beyond the reach of graded roads, sewer lines, and municipal waste removal. But it was actually one of the first cities on the continent to have a master plan, conducted by Sir Alexander Gibbs in 1949.[29] Now on its second master plan, most people nevertheless saw the city as an eyesore, mired by an absence of forethought. Intractable, Dar was quickly expanding beyond the boundaries of its own map in ways that seemed unbecoming of the new nation's capital. Another letter writer commented that Dar es Salaam was trapped by its "arabic shape" that had allowed the proliferation of "shanties." It was time to move instead to a "spacious place" and start "our new and modern capital" undetermined by past colonial occupations and thus truly "African."[30]

Despite Tanzanians weighing in on whether the capital should be relocated, the decision was ultimately put to a party vote in 1972. The decision was an "unusually open decision" in Tanzanian politics and not entirely orchestrated from the center, perhaps in part because it would have such immediate effects on the lives of a broad swath of Tanzania's political elite.[31] Thus, the discussion and ultimate vote was opened up to all party branches and regional councils, though it was always clear what Nyerere wanted.

When the votes were counted, the move to Dodoma was announced the following year. As TANU began making plans for this monumental transition they also hoped to convince the public of the virtues of this expensive relocation through similarly comparing the aspirations for the new capital with the complicated reality of Dar es Salaam.[32] As Dodoma became an environment and place ripe for transformation and the future, Dar es Salaam would recede into a colonial past that the new capital would help Tanzanians forget. The president suggested that taking only the near future into account would miss the fact that Dar faced bleak prospects: "In 20 or 30 years' time the present buildings in Dar es Salaam will be worn out compared with other developments which will take place by then. At that time, too, if we are asked to point at the buildings in Dar es Salaam which made us hesitant to build a new capital elsewhere, people will laugh . . . but in 20 or 30 or 100 years from now Dodoma will be in the centre of the country and Dar es Salaam on the Periphery."[33] The move was predicated on a transcendent future; any less would not be a future worth planning for.

Counterintuitively, the profound struggle that the Dodoma region's history and climate posed also seemed to make it more suitable. In promoting the location, the state argued that Dodoma's "dry and barren" landscape would make it a "good example": "the inhabitants for the most part of this region have tried to work almost against nature and far beyond their power of sweat." Building the new capital in a place where hard work was so essential would prove to "every leader" the true transformative capabilities of TANU.[34] The region's hardships became its recommendation. Dodoma's place in the middle of a drought-prone region would prove the transformative power of ujamaa to remake nature as well as nation. Additionally, all that marked Dar as a problematic capital with a past shaped by foreigners could be averted in Dodoma. This resonated with a global planning zeitgeist that was turning its back on old cities as unsalvageable, eager to start from scratch.

However, as some scholars have pointed out, there was little to suggest that in reality, the relocation of the capital would in fact transform surrounding rural life and rewrite a history of neglect. As Simeon Mesaki and August Nimtz point out, many areas to the north and south of Dar had remained quite poor, despite the old capital's rapid growth; there was no guarantee that rural regions would benefit by the proximity of a new and important city.[35] There was also good reason to point out that while Dodoma might be central, communication with "the masses" would not necessarily be easier. If one considers infrastructures rather than geography, Dodoma was actually quite distant. Mesaki and Nimtz also suggest that the relocation of the capital was never simply an ideological decision to rewrite Tanzania's uneven development but was also a way to shore up state power. In 1971 and 1972, Tanzania sat quite literally between two dramatic political events: a coup d'etat that unseated Nyerere's ally, president Milton Obote of Uganda, and the assassination of the president of Zanzibar (who was also the vice-president of Tanzania), Abeid Karume. Dar in this way represented not just a colonial past but a potentially volatile future, shaped by an increasingly radicalized student body at the University of Dar es Salaam, a wave of workers' strikes in the city, and a fear that Dar was too "obvious of a site" for what the prime minister Rashidi Kawawa called "deliberate provocations made by the enemies of Tanzania to wreck her economy and deliberately provoke workers to destroy tools or resort to strikes"[36] Thus the reasons were complex and reflected a fraught relationship between TANU and Dar es Salaam as much as an aspirational vision of Dodoma remade at the center of the new nation.

Urban Crisis: Reston, Virginia, and New Delhi, India

As plans for the new capital took shape, Tanzania developed a curious relationship to the midcentury suburbs of North America. This unlikely connection illuminates how concerns over an unfolding urban crisis traveled from their origins in American and European cities to influence the plans of postcolonial cities. Embroiled in bankruptcy, racial unrest, and aging infrastructures, postwar American cities suffered gravely from an epidemic of white flight that marked a broader turning-away from urban life.[37] Driven by cynicism and fear, as well as a good deal of utopianism, many mostly white communities emerged across the United States as part of the New Town movement. Seeking to start over in the suburbs, the New Town movement hoped that through excision from the "urban disease," community life could be reimagined and rekindled.[38] Reston, Virginia, remains one of the most famous examples of the New Town movement. This suburb of DC was developed by Robert Simon, who commissioned an urban plan in 1963 from the architect James Rossant. Simon's vision, executed by Rossant, was "to build Reston as a series of dense village centers that just happened to be in the middle of the countryside," and "each would have its own architectural style and a central plaza with shops and things to do."[39] In contrast to many other suburbs driven by postwar car culture, Rossant designed Reston to be walkable by including a concentrated urban development that would also "preserve more of the surrounding woodlands."[40]

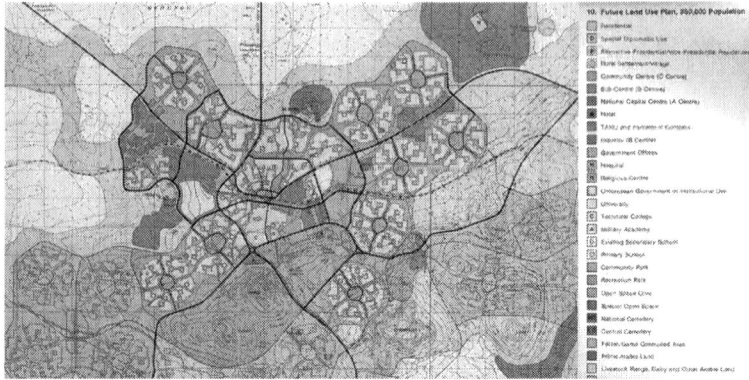

1.3. Future Land Use Plan, Dodoma Master Plan. Originally from Dodoma
master plan by Project Planning Associates Ltd, 1976. Source: Sophie van
Ginneken, "The Burden of Being Planned: How How African Cities Can
Learn from Experiments of the Past: New Town Dodoma, Tanzania, *International New Town Institute*, http://www.newtowninstitute.org/spip.php?arti
cle1050.

While Rossant remains most famous for his plans for Reston, another hallmark of his career was his blueprint for Dodoma's National Capital
Center, a multiuse area at the heart of the new planned capital. Drawn up
nearly ten years after the Reston plan, Rossant's blueprints sit at the center of
Dodoma's master plan, produced by the Canadian planner Macklin Hancock
and his company, Project Planning Associates.[41] This master plan imagined
Dodoma's rebirth as a city "composed of five self-sustaining villages of some
30,000 people each," plans for the new capital bore an uncanny resemblance
to Rossant's suburban DC idyll with each neighborhood displaying its own
"individualistic architecture" that "will enhance and enshrine traditional African values. Each housing unit is to have its own vegetable garden so urban
life can maintain rural ties."[42] Sold to the public as a distinctly African capital, the master plan for Dodoma carried the ghostly blueprint of a suburban
American town built in the middle of a woodland during a time of deep anxiety that cities were unlivable. While the roots of skepticism over urban life
were markedly different for postwar America than they were for new leaders
of Tanzania, Rossant's plans suddenly rendered them deeply connected.

The genealogies of Reston and Dodoma also intersect in ways that reach
back beyond either of their blueprints. As historian Andrew Friedman points
out, Rossant and his firm's partners, including Albert Mayer, had spent the
1940s and 1950s working abroad, honing their town planning skills in the colonial and postcolonial world. In India, Mayer in particular refined his vision
of the urban neighborhood around the concept of the village as the ideal "unit
of production, both economic and creative."[43] He had brought this idealism

about the village to his work on a master plan for Delhi not long before the Reston project began. His work in Delhi was focused on decongesting the city center and creating neighborhoods built around an ethos of face-to-face interaction. Since new urban residents were arriving from villages, Mayer argued that the village must be represented in the concept of the neighborhood. Built upon a paternalistic notion that Indians experienced an atavistic pull to their rural homes, Mayer's work sought to reshape cities in both the decolonizing world and in the United States around a nostalgia for the village. "Both nostalgically and theoretically their hearts are in the villages, and it is a sort of article of faith that the villages are vastly superior to and more ethically habitable than the city. The city must be tolerated, but there is, generally speaking, no creative concept or sense of urgency or sense of identification."[44]

Because of the transnational careers of men like Mayer and Rossant, who traveled between the "developed" and "underdeveloped" world, a surprising consensus of form emerged across quite disparate urban settings. The village became a boundary object that attracted diverse supporters through its nature as both an elastic and concrete form.[45] The village's precise intellectual or cultural antecedents mattered less than its ability to resonate across contexts and gather broad approval. With astounding continuity across capitalist and socialist states, nongovernmental organizations, and multinational donor agencies, the model of the village became an urban salve, whether one was escaping a city for the country or facing the supposed alienation of urban life for the first time. In this context of a globally circulating network of planning and expertise, Dodoma emerged on paper as the new "chief village in a nation of villages," while Dar remained a palimpsest of successive colonial occupations that had relegated most urban Tanzanians to inhabit the margins of their own rich urban past.[46]

From India by way of Reston, the plans for creating a new capital in Dodoma signaled an opportunity to rewrite the relationship between rural Tanzania and its cities. Clement George Kahama, the director of the Capital Development Authority (CDA) responsible for the Dodoma project, noted, "The way God arranged things, Tanzania was settled around the edges . . . but with Dodoma we will have a new centre."[47] As the CDA began the move to Dodoma they continued to work with the Canadian consulting firm Project Planning Associates (PPA) to helm the transition. PPA's master plan for the new city was careful to articulate a vision of Dodoma that positioned the new capital as the linchpin to the nation's ujamaa future, even including a copy of the Arusha Declaration in the appendix to the master plan. Regardless of its execution, the multivolume master plan now existed as the state's platonic ideal of a new capital at the nation's center. Indeed, in light of the absence of any state articulation of urban ujamaa, these plans offer a unique glimpse into what the state imagined an ideal socialist African city to be.

1.4. Sketch for open space and new housing in the Dodoma Master Plan. Originally from Capital Development Authority Archives, Dodoma. Source: Sophie van Ginneken, "The Burden of Being Planned: How How African Cities Can Learn from Experiments of the Past: New Town Dodoma, Tanzania, *International New Town Institute*, http://www.newtowninstitute.org/spip.php? article1050.

Somewhat ironically, these transnational plans that relied on an immense amount of international aid drew up Dodoma as the self-reliant capital of the new independent nation. The city would serve local communities rather than be dependent on global flows. Central to that vision was the fact that the new capital would emerge on the heels of an intensive program of villagization in the region known as Operation Dodoma, begun two years earlier. Presidential planning teams descended on the region, identifying new ujamaa village locations and swiftly preparing plans.[48] Residing not just at the center of the country but at the heart of villagization efforts meant that "the problems and progress of neighbouring villages will be daily in evidence" in the new capital.[49] This embeddedness was also key to selling the expensive move of the capital to both Tanzanians and foreign donors, although some donors, such as the World Bank, remained skeptical.

Project Planning Associates drew out these rural connections in their master plan by invoking the aesthetic of the village in its layout and how it described the new city.[50] By emphasizing the value of village life, the plan offered up a contrasting vision of urban life drawn likely not from Tanzanian experience but from the urban crisis of northern cities. Just as Mayer approached his new plan for Delhi by seeking to nurture face-to-face contact between residents, Nyerere reasoned that the new layout would "give emphasis to [a] community type of living so that when a person comes back from work he doesn't live in isolation and separation."[51] These were not necessarily issues facing most urban Tanzanians but rather suggest the anticipation of an urban future where citydwellers suffered the same problems plaguing communities in the industrial North. But not all forms of communal intimacy were encouraged. Nyerere wanted close urban contact among neighbors but dismissed skyscrapers as the sort of "dense urban environment" that would "be alien to national lifestyles and philosophies."[52] Instead, "a house should not be built so close to another that a chicken from one yard can lay an egg in a neighbor's yard, nor so far away that a child cannot shout to the yard of his neighbor." The president assumed that Tanzanians would feel innately dislodged in a city that did not resemble a village, which ironically echoed many of the anxieties of colonial planners and administrators who assumed Africans would be alienated and "detribalized" if they moved to urban areas.

Planners also rendered Dodoma's stark landscape a verdant agricultural region on paper. New houses would emerge in the plans among the natural topography of the region using hills and valleys to variegate the landscape and preserve "access to vegetable gardens, fruit trees and flowers."[53] To execute this arboreal vision, the CDA began planting trees to fill in the green circles on the new map. Two years later they had planted more than 335,000 plants and trees on 150 hectares as well as developed a vineyard and established two nurseries. But these ambitions to green the city were again more surface than substance. Most notably, creating green spaces to evoke a village aesthetic was at odds with how villagers actually used open spaces for cultivation and animal husbandry. Urban authorities began cordoning off spaces set aside for reforestation to protect them from grazing animals and farming projects, making these spaces not very "open" at all. Dodoma's plans prioritized parks and trees as an expression of its continuity with the larger region rather than the self-help urbanism that inherently mixed strategies of the city and the country but did not necessarily "beautify" city spaces.[54] Thus, while gesturing to the rural, Dodoma's plan did not actually leave room for Tanzania's poor.

Despite the state's intentions to address the spatial inequalities that plagued Tanzania and particularly rural development, the plans for the new capital represented a much higher division between town and country and severely limited where and how citizens could build in the city. The average cost of housing according to the plan was also far higher in Dodoma than in Dar es Salaam. In glaring tension with the plan's village reference point, the capital plan accommodated no self-built housing nor did it allow for upgrading. Even though most residents in the Dodoma region could not afford to build otherwise, city officials began destroying all self-built housing, posting soldiers on the outskirts of the city to disassemble any ongoing self-help construction.[55] The cost of extending essential facilities to households in Dodoma would be ten times the cost in Dar. The Canadian planners had set "exceptionally high, international standards" for the new capitals' "built environment," requiring house materials and building infrastructures that most Tanzanians could never obtain.[56] What seemed to evoke the village on paper was in fact far removed from actual rural realities. There was also little to suggest that in reality the relocation of the capital would in fact transform surrounding rural life.

By 1980, as housing costs remained high, the state initiated a two-year contract with the British construction company Wimpey to construct one thousand units of housing in Dodoma. The company was known for its fast and economic construction (reportedly building a quarter of a million homes per year in Great Britain).[57] But John Sankey, the British high commissioner of Tanzania in 1982, recalled the Wimpey contract as the debacle of "the million-pound house": "The idea was that Wimpey would send out some experts to see the local conditions, design a model low-cost house and then build as many houses as they could within a total budget of a million pounds. By the time expensive consultants had flown out to Tanzania and been provided with houses, cars, servants, education allowances and so on, they were only able to design and build one low cost house before the million pounds ran out. This was the most expensive low cost house in Tanzania."[58] Clearly, many of the tactics that planners used to imagine a rural capital were in fact prohibitively expensive endeavors out of reach for Tanzania. These expenses also represented one of the fundamental frustrations of postcolonial development: the skeletal infrastructures, frequently made "on the cheap" and left behind after colonial rule were an expensive legacy to reengineer. The historical reality of colonial rule made it cheaper to continue building in places already connected to flows of capital and resources, like Dar es Salaam.

It is easy to assess the Dodoma plan as a classic case of poor planning. Patrick McAuslan, a lawyer hired by the CDA to aid in implementing the master plan, realized that for all the plan's references to creating a capital entrenched in Tanzanian culture, the Canadian planners remained in Toronto for the majority of the project and worked under the assumption that they were "planning by bulldozer."[59] With such latitude, the planners were able to ban all building in the city from 1972 to 1975 when the master plan was being developed, with no legal authority to actually do so. Yet despite these attempts to physically erase the existing city, the future was stuck emerging among the still-lurking past. McAuslan was "struck by the extreme oddity of having to draft regulations under the authority of an English colonial ordinance, promulgated in 1956 and based on English models from the 30s, to implement a Master Plan drawn up by Canadian planning consultants and geared, for all its language, to create a North American city beautiful in the heart of the independent socialist self-help society of Tanzania in the 1970s."[60] Indeed, postcolonial cities were frequently stuck within the contradiction that they were originally composed of skeletal and extractive infrastructures while subsequently entangled within thick histories of planning. Furthermore, as synecdoches for the new nation, the blueprints for planned capitals like Dodoma were heavy with expectations of reinvention and transformation.

In observing this gap between plan and reality, this distance between the two represents something other than mere miscalculation. Rather, it captures the very ambitions of Dodoma as a modernist city-from-scratch. James Holston notes in his work on the planned capital of Brasília that Third World governments from across the political spectrum were drawn to modernism "as an aesthetic of erasure and re-inscription." Modernization became a tool of for erasing the past. Executing such plans was nearly a rite of passage to signal the "negation of existing conditions." And as a result, the "utopian difference between the two is precisely the project's premise." But to ultimately realize this new vision, states, bureaucrats, and communities were stuck grappling with the very conditions the plan sought so succinctly to eradicate.[61] Indeed, the gap between the plan and reality was precisely the point *and* the problem of Dodoma's new blueprint.

Bureaucratic Decentralization

As Dodoma's future seemed bright, Dar faced the reverberations of another transformative policy decision: bureaucratic decentralization. Several African countries adopted policies of decentralization by the 1980s—frequently at the behest of the World Bank or the International Monetary Fund (IMF), both of which determined that the process of diffusing centralized power and decision making created a more accountable form of governance as well as a bulwark against state corruption. Tanzania was one of the first countries to do so, years before it became standard procedure. As Nyerere explained when announcing the Decentralization of Government Administration Act, "Our country is too large for the people at the center in Dar es Salaam always to understand problems or sense their urgency."[62] It was, somewhat ironically, an admission that planning from afar frequently failed to take into account local circumstance and contingencies, even as Canadian planners drew up the Dodoma plan and the American consulting firm McKinsey and Company was hired to draw up plans for decentralization.

Decentralization aimed to enable local communities to govern their own affairs, reducing the need to seek approval from ministries in Dar, which Nyerere characterized as both geographically and ideologically distant from rural development. In theory, decentralization would help realize a central tenet of ujamaa: communities should develop through consensus and participation of all citizens.[63] This new geographical arrangement would spread out bureaucratic manpower by relocating planners and bureaucrats outside of Dar and in the regions they served. Development officers who had once reported to Dar would now seek approval from local district and regional councils, with the hope that this would ease any bureaucratic holdups for unfolding villagization projects. This shift in the structure of government had immense and exciting implications for the whole nation. Yet despite its democratizing aspirations, the legacy of decentralization was quite the opposite. While new local governing bodies of bureaucrats moved to the regional level, they remained removed from local politics. Party-appointed bureaucrats whose careers were decided by party elites took over the governance of regional development directorates, and in the process they dissolved councils of locally elected officials.[64]

For Dar, the effects of decentralization were particularly devastating. Not only did it relocate a cadre of experts from Dar to regional centers but Dar's city council, like all urban councils across the country, was abolished following decentralization. The responsibilities of the council were distributed among district and regional offices, and the city was even stripped of its status as a municipality.[65] It was subsequently divided into three districts: Temeke, Ilala, and Kinondoni, each administered by its own commissioner who was beholden to the regional party commissioner. Each district comprised both urban and rural areas and no single district held jurisdiction over the whole city below the regional level.[66] To equalize development space across the nation, rural improvement initiatives were explicitly prioritized over urban ones in each region. The budgets for all major city services, including road building, garbage collection, water delivery, and waste removal, were essentially frozen in time even as the city continued to grow. In a developmental state where all acts of planning are also decisions of what must go unfunded, TANU decided that the old colonial capital had been prioritized long enough. It would be predominantly through the patchwork provisioning of foreign aid projects that the city would receive garbage trucks, city buses, and spare parts, or be able to undertake road building and other infrastructural improvements.

With the elimination of its city council, the departure of bureaucrats, and the concerted effort to focus on rural development, Dar's built environment would be shaped for the rest of the socialist era predominantly through two means: the labor, planning, and provisioning of families and the implementation of regional policy directives. As Goran Hyden commented about decentralization, "the principles of participation and efficiency which featured so prominently in the original policy statements about decentralization . . . gradually receded into secondary positions. Instead, 'operations' and 'campaigns' . . . in which all available resources have been mobilized for a single cause, . . . developed into the most important mode of policy-making."[67] Passed down from high-ranking party officials or Nyerere himself, policy directives focused a flurry of activity and compelled the concentration of labor and resources of local communities on single issues such as sanitation or urban farming. Lasting from a few weeks to in some cases a few years, these directives sought to organize "self-help" efforts to fix municipal problems in lieu of capital investment for fixing urban infrastructures.

In concert with these bureaucratic changes, an archipelago of ujamaa villages began to take shape in Dar's rural areas following the beginning of forced villagization in 1973. This regional push for villagization was known as Operation Pwani. It was the first of several operations aimed at relocating unemployed urbanites into the surrounding region, bringing the village and the city into further tension as well as collaboration.[68] While planning experts worked to fashion Dodoma into the platonic ideal of the village, officials still policed and controlled how and where people could build. Meanwhile in Dar, the shifting circumstances of decentralization drove the city to become even more deeply entangled with its own expanding village periphery, and ultimately dependent on it. Thus, in contrast to the new capital, official planning was essentially abandoned.

So, then, why did people keep moving to Dar? By the 1970s the city was growing far more rapidly than it had in the past. Both men and women were moving to the city in search of work. And while this book is not a reappraisal of the reasons people came to the city, it is still important to consider here in this first chapter, why, when the state was so invested in decentering Dar, Tanzanians continued to migrate to the city at record rates.

The rapid expansion of Dar's population most notably begins after the Second World War. From 1900 to 1948, the city's population increased by around fifty thousand from a population of twenty thousand at the turn of the century. This was actually quite a low growth rate of around 1–2 percent when taken over the fifty-year period.[69] But after the war, as was the case across the continent, Africans began moving to cities in record numbers. Dar faced a housing crisis and the price of food skyrocketed. In the next twenty-three years (1948–1971), the population quadrupled, no doubt sparking postcolonial state concerns that the already-developed colonial capital was becoming the receptacle of too much development attention and a vortex of wasted rural labor.[70] In this way, the state also cared about *who* was arriving in the city: was it men and women who were finding wage labor and ultimately establishing secure lives in the city? Or was it migrants who languished visibly—drinking tea and playing bao—in the urban landscape in between short stints of mostly informal work? As the following chapters demonstrate, both the colonial and postcolonial state saw one's relationship to wage labor as the most important determinant of a person's security in the city. In the state's view, this relationship to labor also determined how someone would engage with and inhabit the city. Finding waged labor particularly at parastatal factories or serving the state as a salaried worker gave you access to different parts of the city, frequently determined what materials you might build your house out of, and gave you greater access to urban infrastructures and to tools of community building and development. The aesthetics of this was no accident: states believed that it was modern workers who would make modern cities. R. H. Sabot notes in his study of urban rural migration in the 1970s that wage earners in particular "come in a sense to embody 'development,' since they are seen to represent all that is progressive, productive and constructive in the economy."[71] While ujamaa valorized unwaged rural development as the ideal form of nation building, the opposite was true for "development" in the city by the late 1960s. Surplus populations in the city undermined rural village development. This also fit into a longer history in which colonial authorities saw casual laborers as unreliable and dangerous to colonial orders, having cultivated alternative economic, kinship, and environmental networks to supplement periods of unemployment by chance or circumstance. These Africans, insufficiently urban, did not need to sell their labor consistently to survive, which made them harder to control.

In 1970, 82 percent of adults in Dar said that they were born elsewhere and yet many of these new migrants were still able to secure wage labor in the city. But as the portion of migrants who arrived with no promise of stable work began to increase, this worried not only the state but researchers across the continent seeing a similar trend. Why were people moving into the city if they were unemployed when they got there? How long would they remain in makeshift housing with little access to urban infrastructure? What economic models explained this behavior and where was the supposed equilibrium point at which reverse migration would occur?

The answer, at least in part, lay in a lack of opportunity elsewhere. Moving to the city was, as anywhere, a calculated gamble reflecting a variety of what geographers called "push" and "pull" factors. Agricultural wages had dropped, ujamaa villages were variably successful and livable, and in the early 1970s rural communities also faced an extended drought. In light of these circumstances, Dar had immense advantages, particularly for those who already lived in the region. Lingering colonial infrastructures meant that Dar remained the "minimum cost location" for both shipping inputs for factories and distributing goods to consumers. By 1966, 75 percent of manufacturing was in the consumer goods industries, as opposed to only 45 percent in Kenya. The biggest market for these goods was Dar, and there was virtually no capital goods production in the new nation (goods used to produce other goods, rather than bought by consumers). Consumer goods industries relied heavily on imported materials and Dar was the biggest port in the country. While the extra distance inland from other locales might seem like a very small part of the total distance these goods traveled, the condition of transportation and roads in Tanzania made it disproportionately expensive.[72] This was a clear and urgent hurdle of economic decolonization and it is also precisely why constructing new housing in Dodoma's capital plan was so much more expensive per unit than building the equivalent homes five hours away along the coast. Only in places where a significant proportion of manufacturing was done with local goods (such as coffee processing) did it make economic sense to put manufacturing plants beyond the coast. Families thus had little incentive to divert their path to the dusty center of the nation with the announcement of the new capital. Not only would jobs be harder to find but consumer goods would also cost more to buy, if and when they were available.

The inherent tension of 1970s Dar es Salaam should be clear by now. The state saw its future as contingent on redressing the uneven development of the past as well as insulating itself from potential invasion or destabilization from both within and abroad. This was manifested in a plan to recenter the nation at its true geographical center while also implementing a policy of administrative decentralization. For Tanzanians, the logic of their lives nevertheless pushed them toward the old capital. But, as the following chapters will show, these new migrants who continued to arrive in Dar ultimately decentered the city in their own way. As the Tanzanian economy began to falter, new and old urbanites alike were forced to turn to the outskirts to build their lives.

Despite their best efforts to turn citizens away from the city's bright lights, Dar es Salaam remained a growing amorphous object in the peripheral vision of a state determined to look toward the rural. These attempts to peripheralize the city are key to understanding by what means and methods Dar's environment would be shaped over the next decade. Sitting on the nation's edge, the state saw Dar as a symbol of past colonial exploitation and its reliance on exploitative foreign relationships. Willed into being at the nation's dry and dusty center, Dodoma would take up a very different position and signal a new relationship with both rural Tanzanians and the surrounding environment.

But if examining the plans for Dodoma helps us understand the state's ideal city as an invocation of the village, efforts to place rural life at the center of the new nation did not diminish the draw of the old colonial capital. Despite what Akin Mabogunje has called an onslaught of "radical spatial policies designed to inhibit the growth of Dar es Salaam," there was, among other obstacles, a nagging power to the colonial orderings that had built the city in the first place.[73] The result of these policies was not the contraction of the city itself but a circumscription of the tools of its growth.

By the 1980s the diagnosis of "urban crisis" was no longer used to describe ailing American or European cities but instead began showing up in headlines and news stories about African cities. It became a cynical and problematic new status quo that continues today. The problems facing African cities have conjured an image of these metropolises as categorically different from urban life in the Northern Hemisphere. And urban crisis is now a term of othering that evokes disorder, poverty, and ecological disaster in cities of the Global South. And yet it is clear that fears of urban life in America at the end of the 1960s helped shape urban policy in cities like Dar. Both TANU and the international development community aspired to a rural future for the new nation. But as the 1970s unfolded many Tanzanians made other plans, keeping the city at the center of their lives.

A key contribution of environmental history has been scholarship that considers how both space and economies are transformed by the emergence of urban hubs. In some cases, like Chicago, this process transforms those cities from outposts to metropoles, but remakes the region's agricultural hinterlands at the same time.[74] In the shadow of these new cities, other would-be hubs linger and sometimes fall off the map as the magnetic pull of a primate city becomes the new all-encompassing economic logic, writ large into the landscape. There is no doubt that Dar es Salaam emerged through quite different historical circumstances than did a city like Chicago. For one, colonialism never transformed Dar's hinterlands on the scale and in the manner of the American heartland. But nevertheless, considering the state's desire to decenter Dar in the 1970s offers an opportunity to examine a prospective second half of these environmental stories of developing hub cities and their transformation of landscapes and economies. What tensions emerged between states, communities, and families as they contended with the process of undoing these path dependencies? How does the logic of urban primacy continue to shape landscapes even as the state in particular seeks to redress the effects of uneven development? What happens, too, when migrants still find more opportunity in the city, while the state withdraws its money for urban infrastructures? If Chicago reveals the scalar process of wheat fields becoming financial futures, perhaps cities like Dar disrupt this myth of the future as a process of ever more abstraction of nature, and brings us back to the ground.

Chapter 2

BELONGINGS

Settling and Unsettling in Dar es Salaam

Outskirts are the state of emergency of a city, the terrain on which incessantly rages the great decisive battle between town and country.

Walter Benjamin

In 1972, as the CDA policed the borders of Dodoma to prevent self-building, Dar's urban authorities announced the end of slum clearance, a policy that had been the hallmark of postcolonial attempts to rein in the city's unplanned residential areas. Indeed, as the state sought to preserve the clean lines of the future planned capital, Dar's built environment became increasingly heterodox, unfolding off the urban map. While the growing illegibility of the city's landscape certainly reflected the new regional and decentralized approach to Dar's governance, it also—perhaps counterintuitively—reflected a colonial legacy of policing the boundaries of the city. This chapter seeks to show how in both the colonial and postcolonial period "unplanned" (or "informal") urban space was never the opposite of the city's planned areas but rather its inescapable counterpart. And as the city's urban crisis worsened over the course of the 1970s, the central "planned city" was increasingly propped up by the "unplanned city" on the outskirts.

Uncovering these sedimentations of colonial rule across the postcolonial urban landscape highlights the stubborn contours of belonging in Dar's built environment.[1] As John Crowley defines them, the politics of belonging are the "'dirty work' of boundary maintenance," and indeed this chapter is about boundaries both literal and figurative.[2] And while the socialist state's language of liberation and "familyhood" was framed in opposition to the alienating boundaries of colonial belonging, Nyerere's policies toward Dar sustained profound continuities with the colonial period. Jobless migrants struggled especially to find spatial belonging in Dar and this process produced certain kinds of urban environments that broke along boundary lines.[3]

In the historiography of "belonging" in Africa, urban spaces have figured prominently as new terrains where rural migrants must renegotiate
their identity, citizenship, and sense of community. This is in part because
colonial powers historically excluded or severely limited Africans' access
to cities, leaving individuals and communities with multiple struggles to
prove their right to the city or find other modes of making place outside of
hegemonic channels of legitimacy. Colonial knowledge production also
defined Africans almost exclusively through a language of ethnicized rural
life, portraying them as existentially out of place in cities, ever unbelonging.[4]
Migrants to the city also had to negotiate their own complicated connections
to their rural homes as well as the new experiences that came with bumping
up against different ethnic groups in the city. As a result, these struggles to
belong have become a major theme of literature on urban Africa.[5] Unlike
much of this work, which focuses on social and cultural dimensions of urban
belonging, this chapter focuses on the material and spatial manifestations
of these contests. As established urban residents, new migrants, wage workers, and members of the "informal economy" negotiated each other and the
state, they also shaped urban environments. Finding belonging in the city
was frequently about finding a place to live but also about securing the very
materials of provisioning for life in the city.

The first two parts of this chapter consider how Dar's boundaries and
neighborhoods were drawn up by colonial and then postcolonial officials to
reify urban space and limit the city's population. Here, the chapter considers
who garnered access to housing and land in the city and what institutions
secured this access. It also considers how those who were excluded made Dar
their home through alternative means. In the third section and the heart of
this chapter, I explore how emerging food shortages in the 1970s along with
a constricting urban budget turned the logic of urban belonging on its head.
Those who had hoped for (or found) inclusion in the planned city and access
to urban infrastructures were now going to ground in the periphery. The state
also urged and frequently forced urbanites to the periphery to farm in order
to alleviate food shortages and depopulate the city. While these measures by
the state may have been attempts to discourage urban migration and keep
the city exclusively the domain of the wage laborer, it in fact had the effect of
transforming and redefining the space and nature of the city itself.

Seeking Entitlement in a Colonial Capital

In both the colonial and independence periods, claiming space in the city
was a constant struggle for inclusion in what James Brennan has called "urban entitlement."[6] This continuity across the divide of decolonization re-
sulted from enduring urban planning laws first implemented by the British
colonial administration after the Second World War.[7] By the interwar period,

housing shortages, long an issue shelved by colonial authorities, had become an urgent matter.[8] The new influx of urban laborers in the 1940s exacerbated this shortcoming of the city. Beyond numbers, colonial officials also sought to mediate the kind of Africans who were allowed permanent place in the city. To do so, urban administrators, town planners, and engineers worked together to cultivate permanent middle-class African enclaves. City officials also permissively allowed extralegal and extraurban settlements on the periphery. Through instruments of town planning, the "distinction between town dwellers and transient inhabitants" became entrenched in both law and landscapes.[9] While in reality these distinctions were never as Manichean as they might have been on paper, the former were integrated into neighborhoods imbued with legal and material markers of permanence and the latter had more freedom from regulations when building but virtually no legal protection or economic assistance. Those beyond the boundaries of entitlement garnered by residence in planned neighborhoods were legally vulnerable to campaigns of urban expulsion, as were those who remained in the city without jobs.

The colonial state constructed these new middle-class enclaves to minimize their own anxieties regarding the increasing presence of Africans in cities and at a moment when the legitimacy of colonial rule was being called into question. They hoped these environments would transform an unrooted migrant male population into settled wage earners with nuclear families.[10] These neighborhoods were the spatial counterparts to a burgeoning field of social science research into how urban Africans lived. J. A. K. Leslie's *A Survey of Dar es Salaam*, a cornerstone text of Tanganyikan urban history, was part of this broader late colonial project and one of many written by social researchers or colonial officials across Africa.[11] Texts like Leslie's were not merely reports on the state of urban Africa but were commissioned by colonial states as tools for bettering cities. The book jacket to Leslie's survey attests: "There is much to be done to improve life in African towns, and this book is a contribution toward the necessary background knowledge."[12] When identifying the need for a survey of the city in the first place, Dar's administrators expressed particular concern that it was "the new migrant" who tended "to be the law breaker and trouble maker."[13] By framing the perennial stream of new migrants as the source of trouble, the colonial state sought to both halt this migration and find ways to ensure they could be properly absorbed into the city. But with rising urban prices, it was not just permanent wage labor that would stabilize this new population but subsidizing their access to "food, housing, consumer goods, and human mobility," since these amenities were believed to produce rooted, nuclear families.[14] The resulting entitlements gave "unprecedented material meaning to urban space" for some Africans, separating "lawbreakers" from those who had earned a right

2.1. Map of Dar es Salaam. J. A. K. Leslie, *A Survey of Dar es Salaam* (Oxford: Oxford University Press, 1963).

to the city through steady employment.[15] As I will demonstrate in the next chapter by examining the discourse surrounding building materials, the colonial state imagined that these sorts of material entitlements would remake rural migrants into well-behaved urbanites.

The African neighborhoods designated for improvement in order to cultivate this middle-class belonging included Ilala, near the center of town; Temeke, in the southern industrial quarter along the railroad tracks and harbor; Kinondoni, to the north; and Magomeni, a western suburban area along Morogoro Road.[16] The development of Magomeni offers a good example of how the state envisioned middle-class African enclaves. The director of town planning imagined that the African suburb would eventually accommodate two hundred thousand persons, offering "all necessary services, shops, schools etc, to give the proper urban background to their lives."[17] And yet since most

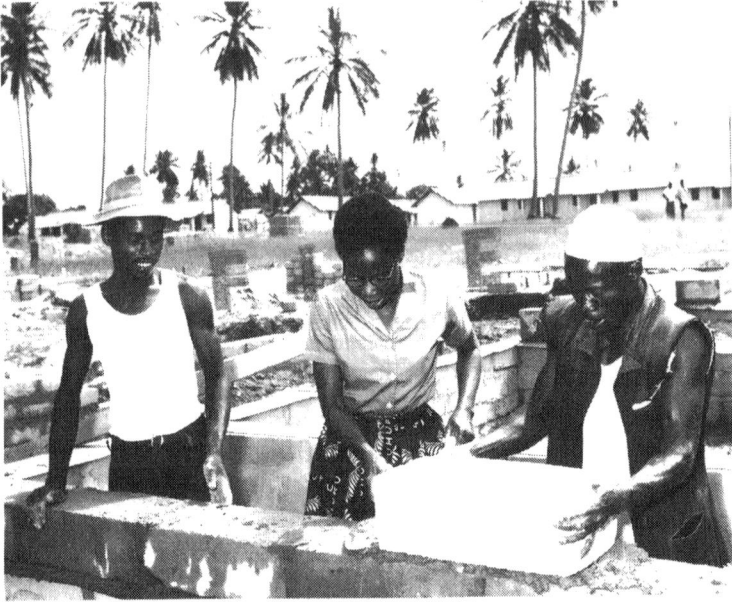

2.2. Men and women building a house with concrete blocks as part of the Magomeni Self-Help Housing Scheme, 1962. Photo courtesy of Tanzanian Information Services.

Africans could not meet food prices on their wages, the government subsidized many necessities for workers to minimize labor unrest.[18] Those who could afford to meet the salary requirements for Magomeni were thus doubly rewarded. Part of the Magomeni scheme was also set aside for residents to self-build, but the quality of construction and materials was highly mediated by urban authorities. Plots for self-building were dispensed only after tallying a score for each applicant that included their income, choice of building materials, family size, and savings. The result was to effectively exclude those who could benefit the most from this middle-ground option. These rejected residents instead looked to informal neighborhoods.[19]

While the colonial government took great care to choose these middle-class residents, the same care was not given to constructing the infrastructure of these new neighborhoods. Despite planning Magomeni to accommodate a community of two hundred thousand, they equipped the neighborhood with only four water spigots. Moreover, city officials never seriously considered constructing a sewerage system. The planners noted, "From the point of view of territorial priority . . . Magomeni's turn for a sewerage system might come up in fifty years." Authorities also remarked that in the Temeke neighborhood, building was permitted with "no form of sanitation whatever. The cost of installing water carried sewerage is hardly worth estimating. Unless this

2.3. Magomeni Self-Help Housing Scheme, 1962. Photo courtesy of Tanzanian Information Services.

condition can be waived the scheme might as well be abandoned straight away. To fulfill the condition it would of course be necessary to have piped water in every single house."[20] Thus, even neighborhoods groomed for permanency and stability lacked basic forms of urban infrastructure, and these services were never deemed part of the material necessities that did form entitlement. From the colonial period onward, successive master plans in 1949, 1968, and 1979 all addressed the dire need of infrastructural expansions and improvements on paper, but few were actually improved due to both lack of funds and the eventual prioritizing of rural development. This deferral of maintenance would come to perennially plague the growing city.

Living beyond the boundary of entitlement in the colonial capital, the rest of the African population settled into the unsurveyed peripheries of the city, where there were older, "informal" neighborhoods that had not undergone slum clearance. For those who ventured farther out, Dar's hinterland could be rocky and sandy, but it also included plantations, farms, and villages.[21] Much of the landscape was also covered in coconut trees. By the interwar period, the town's outskirts were still sparsely populated. Some of this land was alienated from local populations and owned through "freehold" tenure.

Africans frequently squatted on this land, conscripted to guard the plantations of coconuts and, in return, live and cultivate on part of the land.[22] Other
than some "crown land" held by the government, a majority of the land outside of these plantations fell within the legal category of "customary rights."
This land was under the guardianship of local communities deemed "native"
to the area rather than individuals, a practice that assumed forms of ethnic
unity that did not always exist. Those with claim to the land frequently sold it
or rented it out, much of it ending up in the hands of Indians and Europeans
by independence.

Settling on the margins during the colonial period was not hard to do as
long as one's material life conveyed both rurality and impermanence. The
periphery of the city was also where those criminalized for unemployment
would go to ground to hide out from the state. Thus it served as both a refuge
and a labor reserve.[23] As AbdouMaliq Simone has written, the periphery of
colonial cities allowed for a distinct form of African urbanism to be possible because "the 'outsides' of the colonial city" were often "the only available
spaces in which the particular experiential wisdom of African urban residents could be enacted. . . . The rural areas sometimes could be places where
the colonial gaze wasn't as strong; where urban Africans didn't always have
to show a certain measure of compliance."[24] In other words, other forms of
belonging became possible outside the boundaries of urban entitlement.
While these areas are frequently considered *not quite* part of the city, there
is good reason to argue that in Dar, as in many African cities, they actually
represent the heart of African urbanism.

While authorities casually sanctioned this use of Dar's hinterlands, they
also signaled their legal right to reclaim this land as needed. In the 1950s
planners created a 9.5-mile zone around the city to act as a "buffer between
the town and the rest of Africa."[25] The chief planning officer wrote that "this
zone of control" would serve as a place for the development of "small and
properly designed satellites either in the form of native villages or of rural
farming settlements . . . to provide food for consumption within the town."[26]
It could also, however, be conscripted at any time into the borders of the city.
This was entrenched in colonial law through the publication of a 1953 circular declaring that any designation of land as a planning area "extinguishes all
customary rights."[27] Furthermore, with the passage of the Town and Country Planning Act in 1956, the minister for lands gained the power to declare
planning areas and then "set in motion the preparation of general and detailed
planning schemes."[28] Urban officials could claim any land that Africans had
built and lived on outside of the designated perimeter of the city in the name
of planning.[29] Those displaced would be compensated based on the value of
improvements on the land. And since residents in these areas were required to
build with "temporary" materials, compensation was always quite low.

While the peri-urban area may have served as a place of retreat and a refuge for articulating a form of urban settlement outside the European image of the city, it was also a constantly unsettled and unsettling boundary. The city's boundary functioned as the labor equivalent of a cordon sanitaire, a buffer zone beyond which residents were outside the segregated landscape of the city but within the orbit of its labor needs. The colonial state could avoid imperatives for more housing by even encouraging this reservoir of unplanned settlements as long as they were seen as "villages," exerting planning control over them when they became something more pernicious. Claiming these areas as unsanitary could also justify their removal.[30] In the coming decades, this peri-urban zone continued to serve as a pressure-release valve for the city's ongoing housing shortages. One colonial municipal council member wrote in 1957 that any administrative officer who came to work in Dar would need "time to adjust himself to the conditions of Dar es Salaam with its large and transient population, its development problems." Likewise, the city itself "is a curious conglomeration of urban, suburban, and semi-rural and completely rural districts."[31]

Perhaps the most crucial legacy of Dar's urban space under colonial rule was the great importance of formal employment. Formal employment garnered families access to certain spaces of middle-class belonging as well as the subsidization of goods and literal belongings. However, African neighborhoods were the densest, had the poorest infrastructural amenities, and were never large enough to accommodate the city's population, even those with formal employment. Thus, many workers remained outside the city's boundary or in unplanned but older neighborhoods. The colonial state tacitly encouraged this because these neighborhoods could later be razed by designating them a "planning area." These laws played both on Africans' precariousness and the unwillingness of the state to offer adequate housing and municipal infrastructure to support a growing population. In this way, these areas are entangled spaces shaped both by the freedom to urbanize in unconstrained ways and by exclusion from physical urban infrastructures. While notions of permanence and planning guided how the colonial administration thought about space in the city, for most Africans what was far more important was access to wages and commodities that they could not get beyond of the orbit of the city. Access to piped water, electricity, and trash removal was secondary.

Finding Space in the Postcolonial City

In the first ten years after independence, Dar's urban policy took shape around the desire to desegregate the raced and classed neighborhoods of the city. Perhaps the key policy of this era was a sustained attempt at "slum clearance." The city's 1968 master plan prioritized eradicating the city's informal neighborhoods—both "embryonic" and "larger squatter areas" and replacing them with twenty-five thousand new houses.[32] But the state's attempts to replace unplanned neighborhoods with new housing orchestrated by the newly formed National Housing Corporation (NHC) failed to even minimally keep up with urban growth. These new housing developments also continued to exclude the same class of people who had been excluded by colonial policies. Dar's "aggressive, well-educated and better paid urban dweller" obtained leases and planned plots while "low-income, rural migrants who lack[ed] formal education and the skills relevant to urban living" were routinely left out.[33] Middle-class Tanzanians (frequently, state employees found housing in new NHC schemes) were also eager to gatekeep their neighborhoods and protect their own entitlements from new arrivals.

The poor were not, however, the only urbanites facing displacement in the name of desegregating and equalizing city spaces. In 1971 the president announced the Acquisition of Buildings Act and the state seized all privately owned urban property valued over one hundred thousand Tanzania shillings.[34] Nyerere argued for the necessity of nationalizing these buildings as an intervention against a rich and politically connected class of Tanzanians owning and controlling rent. The act was specifically aimed at limiting the ability of politicians to become absentee landlords in the city and accumulate wealth through political connections.[35] The effects of nationalization, however, fell disproportionately on Tanzania's Asian population, who also disproportionately represented the property-owning class in the city; all but 2 percent of the property nationalized under the act was owned by people of Asian descent. In the next few years, thousands of Asian Tanzanians left the country feeling betrayed by the state and convinced they no longer belonged.[36]

Nationalizing buildings virtually stopped commercial and residential investment in the city. For example, Kariakoo, the commercial center of the city, was slated for redevelopment in both the 1968 and 1979 master plans but in neither instance did much change. Beyond the constraints of the Acquisitions of Buildings Act, redevelopment failed because over the course of the 1970s, Dar also suffered from a severe shortage of building materials, which I discuss in the next chapter. These shortages led to racketeering and supply hoarding. When building materials were scarce, it took "more than five years" in some cases, like the Mount Usambara Hotel, to finish construction on new buildings.[37] The Acquisitions of Buildings Act also dramatically proscribed the stewardship of existing structures since owners feared that adding too much value to a building would lead to its confiscation. All the buildings that were nationalized were maintained by the overwhelmed and financially strapped NHC. Dense, multistory buildings became nearly nonexistent for the rest of the socialist period unless they were built by the state.

These political and material circumstances meant that the city would expand outward rather than upward, through the building of small, unattached one-story homes built by urban residents known as "Swahili-style" houses.[38]

A year after the Acquisitions Act, the state announced plans for administrative decentralization and Dar was stripped of its municipal status. With the shifting focus to rural development, the ongoing project of slum clearance was now considered "too costly and socially disruptive"; the 1968 master plan was also effectively canceled for reasons of cost as well.[39] In its stead, Tanzania became one of the first African countries to adopt a policy of upgrading "slums" rather than demolishing them. While this was a prescient moment of realism regarding the financial limitations of both the municipality and urbanites to afford "formal" housing, the policy did not signal a broader state acceptance of unchecked growth of the city, nor did it reflect a recalibration of what materially constituted "formal" or "planned" housing.

The state partnered with the World Bank to improve the infrastructure of unplanned areas and encouraged residents to apply for a loan from the newly created Tanzanian Housing Bank (THB) to build their own houses.[40] With a loan in hand, a family was put on a waiting list for an official plot where they could then build a structure that met city building standards. Despite good intentions, the program was severely crippled by the maze of paperwork and fees involved to obtain a plot. Simply in official fees, obtaining a legal right of occupancy was equal to a few months' salary for a full-time worker and took on average 280 days and included twelve different steps.[41] Self-employed applicants had to produce two years' worth of monthly income receipts to be considered.[42] And for those who did apply and were approved, it could still take four years to get a building permit. Indeed, what had not changed despite the radically new policy was the imposition of standards that most Tanzanians could simply not meet—whether those standards were financial requirements or building codes. By the 1970s, as wages shrank and a growing proportion of the population did not hold jobs in the formal sector, fewer and fewer urbanites could gain housing through these programs.

And yet, while shrinking wages made formal work less lucrative, parastatal factories and state agencies frequently doubled as conduits for urban belonging. The city's major industrial labor sites offered a variety of community services for their employees, creating their own nodes and networks of development. Since labor legitimized one's presence in the city, access to urban space, education, and resources was a perk of earning a wage. Major parastatals frequently provided housing for their workers, organized building cooperatives, or offered educational opportunities. The factory in particular came to be the urban equivalent to the ujamaa village: a place of collective production deemed essential to nation building as well as personal development. In fact, any factory or "economic institution" with more than fifty employees who were TANU party members automatically doubled as a party branch office. To name a few, this included the Tanganyika Packers meatpacking factory, the government press, and two textile mills, Friendship Textile Mill and Tasani.[43]

The Friendship Textile Mill—built in 1965 by the Chinese government for Tanzania—offers an illustrative example how housing, belonging, urban development, and labor were entangled in the city. With 4,328 workers, the Friendship Textile Mill, also known as Rafiki, was the largest factory work-force in East Africa, touted as a "triumph for appropriate technology" as it provided essential textile goods to the new nation and lessened the demand for imports.[44] Rafiki offered living quarters for five hundred families in ad-dition to community resources such as a nursery, a literacy program, and secondary school where workers could take courses. In fact, the mill even sent 153 employees abroad for further education.[45] The factory's canteen pro-vided subsidized meals and a number of women's clubs taught crafts such as knitting, tailoring, basketry, and gardening as a means to earn supplemental incomes.[46] The factory even had its own jazz band, which was not uncom-mon among parastatals and government agencies.[47] These house bands were paid salaries and garnered benefits such as pensions and health care. The Tanganyika Textiles Corporation also arranged employees into cooperative ventures outside of their factory work. For example, the factory bought em-ployees a boat to start a "fishing group" and trained workers to raise poultry for extra income.[48]

Parastatal workers who were not provided with housing on site could still organize through their factories to form building cooperatives.[49] Designed to pool experience, muscle, and money for home construction, building co-operatives emerged as a way to help poor Tanzanians avoid the costly "'third factor' of hiring contractors or builders." With a community at hand, each future home owner "does not pay any penny, he/she can either prepare local brew or slaughter a goat/cow/fishes which they eat and as a thanks to the participants."[50] As building materials like cement became hard to find in the late 1970s, these cooperatives could also provide a way to secure materials in bulk from factories and avoid the steep markup of black market prices.

The first cooperative society to form sought land for building in the peri-urban neighborhood of Mwenge in 1971 and formal employment was a requirement of joining the Mwenge Housing Co-operative Society. Aided by special access to plots and loans, the cooperative planned to construct a neighborhood with roads and storm drainage, as well as houses, a coopera-tive shop, a social center, a bar, and a clinic without any help from the munici-pality to provide these urban infrastructures. The neighborhood became one of the few outside of Dar's center served with a sewerage system.

And yet despite state encouragement, as more cooperatives formed they frequently confronted the same barriers individual builders did in attempting to get loans or titles from local land offices.[51] The Sigara Cooperative, for example, consisting of 167 workers at the Tanzania Cigarette Company, applied for plots for each member but were granted only eighty. Once they were assigned plots, these members were obliged to apply for a loan from the Tanzanian Housing Bank to cover construction costs, and the bank offered only forty of those society members loans.[52] Perhaps most ironically, the Makaki Housing Cooperative, formed by employees of Dar's own housing department, had banded together because they couldn't otherwise find space in planned neighborhoods. They too never received title deeds to their plots. And yet despite these obstacles, when housing cooperatives did work they aided urbanites in not just building homes but extending the infrastructure of the city. Similar to the subsidies of the colonial period, formal labor provided urbanites with more than just an income but also access to development projects, housing, and material security in the city. These projects and cooperatives, though, also reveal how dramatically Dar's regional government (once the municipal government was disbanded) had retreated from funding and executing their own infrastructural projects.

Despite their relative privilege, many wage earners never got access to a surveyed plot or to building loans. By 1975 Manzese, one of Dar's largest unplanned neighborhoods, housed about ninety thousand people, making the neighborhood larger than Tanga, the nation's second-largest city.[53] One-third of Manzese's population were full-time salaried employees, with 22 percent working in manufacturing and 17 percent working for the government.[54] State employees, party members, and newly arrived unemployed migrants all made up these informal neighborhoods. By the mid-1970s many of Dar's residents had also cultivated a general ambivalence to whether they lived in "planned" or "unplanned" areas.[55] And in fact, when the city began facing routine food shortages beginning in 1974, families realized that there were many practical reasons to avoid planned areas where there was less room for improvising space to meet the new needs of urban life through gardening and keeping livestock. And while the state could theoretically remove unplanned neighborhoods in the name of planning, it was not politically nor economically viable for the socialist state to do so.[56]

These unplanned neighborhoods emerged among the creeks, valleys, hills, and mangroves of Dar's transforming borderland. Ideally, early builders in new areas might latch onto one side of the city's main arterial roads for access to town. Houses along that road would frequently develop first and fastest while subsequent builders would settle along narrow, sandy pathways connecting houses to each other and the main road. These paths might be no wider than the space necessary for a handcart to pass by, selling water, col-

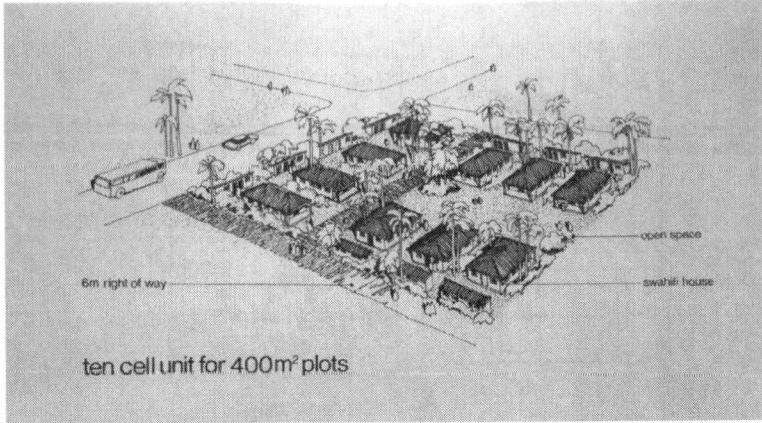

2.4. "Ten-cell unit sketch" Marshall, Macklin and Monaghan, Ltd., *Dar es Salaam Master Plan Technical Supplement*, 1979.

lecting trash, or peddling firewood and charcoal. Garbage trucks and other urban maintenance vehicles could not gain access. About 15 percent of residents in these new neighborhoods did not have regular access to urban water sources and instead drew water from local wells and stream beds. Only about 17 percent of urbanites accessed water through private connections, mostly through drilling their own boreholes rather than being attached to the city water system.[57] The vast majority relied on sparse public standpipes where water was freely dispensed but could also be vectors for waterborne diseases. These neighborhoods were also not hooked up to sewerage pipes but instead relied on outhouses. By 1975 half of the city lived in "unplanned" neighborhoods, and by 1979, when Dar's population was estimated at 794,000 people, this percentage had climbed to two-thirds.[58]

The Ten Cell and Governance by Directive

Planned or unplanned, all urban space had in theory at least two things in common: no one lived beyond the reach of development directives passed down from regional authorities nor outside the most intimate form of postcolonial governance, the ten cell unit. With the virtual end of capital investment in urban infrastructure after decentralization, the ten cell and the policy directive became the modes by which the state commanded communities to develop themselves and enforce their own boundary lines of belonging. Regardless of how you acquired land—whether you were in a middle-class neighborhood or an unplanned squatter settlement—whether you lived in a rural ujamaa village or an *mtaa* (neighborhood) in the middle of Dar es Salaam, all Tanzanians were in theory united by their shared membership in this core socialist configuration of the ten cell.[59] Every *kaya*, or household, when grouped with nine other households in the same area, formed a ten cell unit (*nyumba ya kumi kumi*), the "basic organ of TANU" that represented "a fundamental extension of party and government at the grassroots level."[60] Descending from the nation to the region, district, ward, and neighborhood, the ten cell was TANU's most local arm of government and most fundamental unit of national belonging. The ten cell merged a philosophy of community participation with an ideal geography for living. In this ideal geography, houses would be clustered close to one another and oriented toward a central shared development space where the ten cell community would come together and meet.

Ten cell units were devised to create a uniform and egalitarian framework for development across the nation while also addressing the specific needs of communities. Ten cell units were regularly convened by their *mbalozi*, a rotating leadership position that drew from the community, and these leaders would then become the avenue for information and directives from the state. By attending TANU meetings, political rallies, and development council meetings, these *wabalozi* would return to their unit and report on newly announced development objectives.[61] For example, with the inauguration of a Health Week in Dar, ten cell leaders distributed wastebins to cell members and ordered house owners to "white wash their houses."[62] Members of every ten cell unit were expected to not only keep their own houses and environments clean but work to improve their shared space by taking on tasks such as clearing grasses from the footpaths, helping with the construction of a cell member's new house, or taking "turns on the farm of a sick member."[63] Checking up on the ten cell, members of TANU's Youth League patrolled the streets of the city to make sure people were complying and "do not throw rubbish on streets and on pavements." *Askaris* (guards) were posted at dumping areas to make sure people were not sorting through the waste.[64] In the absence of sanitation infrastructures to serve the expanding city, the state hoped to instill an "environmentality" in Dar's residents to monitor their own space that could be implemented through this infrastructure of belonging that invoked both the neighborhood and the nation.[65]

As an increasing proportion of Dar's neighborhoods were established informally, ten cell leaders and mtaa offices also served as repositories of land sales enacted through customary law. Unsurveyed and unserviced land that locals sold to new migrants (or frequently also their own family members so they could accumulate more land) was subdivided with the aid of wabalozi. The *mzee* (elder) from the region who held customary land rights and the new resident would negotiate a price and then walk out and mark the property line with the ten cell leader or other local elder who would serve as a witness. In the nearby town of Morogoro, as people began settling into the peri-urban zone in the 1970s and taking over the old Kihonda sisal plantation, local village officials worked with a "barefooted surveyor," someone trained to use a sisal rope and bush poles to mark out newly subdivided boundaries.[66] One resident of Dar's Makongo Juu neighborhood explained to me that when he bought the land to build his house, no formal legal documents existed for years to acknowledge his ownership. But he did have an informal record of the land sale signed by himself and the mzee, which was kept at the local mtaa office for safekeeping and reference.[67] These networks of local governance also served to adjudicate disputes when they did arise. Through these processes of both formal and informal authority, the ten cell unit became the arch-infrastructure of the decentered city, dispensing land, ordering the sanitary upkeep of neighborhoods, and addressing communal issues.

But beyond these tasks of community maintenance, ten cells also became the local tool for much larger state "operations" and environmental transformations as first food and then more general commodity shortages became pervasive. Announced in newspapers and over the radio, citizens would also get word of such "operations" through the network of the ten cell: "He [Nyerere] had a unique communication system. He had authority but he was imbued with populism and the will of the people as well. He would call the elders of Dar es Salaam to, say, Mnazi Moja [a large grounds for public gatherings] and make a major announcement so that even your grandmother would understand the importance of food security, for example. And then regional managers would then want to be number one and so on down from there to the cell level."[68] With this system, news residents might have otherwise avoided if it were transmitted over the airwaves was now delivered to them by neighbors or even family members. Everyone was part of a deeply paternalistic system that could both call on citizen action and enact regular duties such as tax collection, enforcing urban farming, or distributing ration cards for ward cooperative food shops when goods were in short supply. But whether Nyerere's calls to community action were always received with the enthusiastic will of the people, as my interviewee suggested, is by the very nature of governance-by-directive harder to historically recover.

The first major operation of this kind was Operation Kilimo cha Kufa na Kupona (Farm for Life or Death), inaugurated by President Nyerere in 1974. Following food shortages and the beginning of a devastating Sahelian drought, Nyerere ordered urban residents to transform all their usable outdoor space into urban farms to alleviate national food shortages. As I will examine in chapter 5, the operation lasted two years and also required factories to start farms in the outskirts of the city. Two years later, Nyerere announced another major policy directive, Operation Kila Mtu Afanye Kazi (Everyone Must Work) in July of 1976. While generally aimed at increasing the city's food supply, this operation was focused particularly on removing all urban unemployed from the city to work on agricultural projects outside the city.

In announcing the campaigns for repatriation to rural areas, Nyerere warned that he did "not want to hear of people staying in towns without work." He also added that people should no longer queue at the labor office while there was "work on the land. There are fifty villages in Dar es Salaam region alone with good land."[69]

The fifty villages to which Nyerere was referring were not historical villages of the Zaramo and Shomvi in the region, but ujamaa villages formed under Operation Pwani, the compulsory villagization order in 1973. Like other compulsory villagization efforts across the nation at the time, Operation Pwani sought to replace the region's sparse settlements with denser villages. For example, in the neighborhood of Nyantira, eighteen kilometers southwest of downtown Dar, indigenous landholders were forced by the state to abandon their homes and join the nearby ujamaa villages of Buza and Kitunda.[70] Regional authorities also consolidated workers on a nearby sisal farm into the peri-urban village of Rangi Tatu. Many resisted these attempts in whatever way they tacitly could. Some inhabited the villages only partially while continuing to travel out every day to perform activities such as tending their farms in other parts of the region, using their private fishing boats, or excavating sand and gravel to sell as building materials. Others moved between the city and the region for work, hoping to evade the dragnet of the authorities.[71]

Building on the infrastructure of Operation Pwani, Operation Kila Mtu Afanye Kazi was an attempt to forcefully implement what had long been implicated in planning laws and urban entitlement: that the unemployed had no right to the city. After a sluggish start, in November of 1976 the regional party secretary Joseph Rwegasira began enforcing urban removals with alarming zeal. In fact, the operation quickly became known as Operation Rwegasira because of his avid enforcement. Because Dar was now administered as three districts rather than a municipality, Rwegasira warned each district that all three had to enact similar bylaws to ensure enforcement of the campaign.[72] At a time of increasingly tight budgets, Rwegasira had a startling ten million shillings earmarked for the resettlement of fifteen thousand of the city's unemployed.[73] These funds helped organize deportation and settlement as "people's militias" swept through the city picking up anyone who could not provide an identity card issued by their ward or ten cell leader that would prove employment and thus belonging in the city. Another way to avoid removal was to brandish a "shamba card" (farm card) that proved that you were already cultivating land in the region. As a result of this slapdash method of identifying supposedly illegal urbanites, the militia frequently just rounded up those on the streets who had no form of identification on them.

Women were especially vulnerable to these sweeps as part of a growing animosity toward women living in cities, particularly if they were unmar-

2.5. Newspaper caption: "Prime Minister Ndugu Rashidi Kawawa participating in preparing new settlements for prospective villagers in Mwongozo
village in Temeke District during the tour of areas in Dar es Salaam region."
Daily News, November 24, 1976.

ried.[74] Marriage and social reproduction rather than a job protected women's
right to the city, but even then they were still vulnerable to getting swept up
in removals.[75] For example, the *Daily News* interviewed Tatu Omari, a woman out buying food for her child when she was picked up. In the aftermath, no
one in her family, including her waiting child, had any clue where she was.[76]
Another woman told the newspaper that she worked at a hotel on Mafia Island (off the coast of Tanzania) and had been picked up while visiting Dar for
a meeting with the Tanzania Tourist Corporation. Rwegasira turned to ten
cell leaders to keep an eye out for women seeking marriages "of convenience
to avoid being resettled in the villages" and to be vigilant for women tricking
authorities by engaging in "embroidery during the day and prostitution at
night."[77] Citizens also wrote to the newspaper to complain that prostitutes
were evading removal from the city by obtaining false identity cards from
local police.[78] Single women were also frequently unable to rent houses or
rooms from the NHC housing stock because it was a "universal" policy of
sublandlords not to rent rooms to unmarried women since the assumption
was that they were prostitutes.[79]

Once rounded up, both men and women were dropped off in regional ujamaa villages such as Geza Ulole, Mwongozo, Tumaini, Yale Yale, and Cheke-
ni. Those dropped off complained of not being given time to even collect any
of their personal belongings. Older villagers already settled in the area would
help by providing "utensils, sleeping mats and bedding to the new settlers un-

2.6. Newspaper caption: "A Tanzania Film Company (TFC) crew shooting a scene for part two of the feature film *Fimbo ya Myonge* ten years later." *Daily News*, January 1985.

til such time as they can afford to buy their own."[80] These villages frequently lacked water sources and building a new home required clearing dense land first. The temporary huts given to the new arrivals resembled, as one resettled villager put it, "one long train with compartments."[81] After establishing their own homes, those who stayed were also required to cultivate at least one hectare of land.[82] Additionally, villagers had to contribute monetarily to TANU and donate funds to local infrastructure projects such as road construction during visits by party officials.[83] Flung outside the boundaries of urban belonging, these men and women were now compelled to participate in regional rural development to secure their places.

To encourage these relocation campaigns, in the middle of the 1970s the Tanzanian state developed and released its only feature-length movie, *Fimbo ya Mnyonge* (The poor man's stick). The film tells the story of a young and lazy rural man who makes his way through life by trying to avoid hard work. His name, Yombayomba, is a play on *ombaomba*, the Kiswahili term for a beggar. One day, Yombayomba leaves his village for Dar es Salaam, where he envisions an easy life of buying and selling goods for a handsome profit. But after he fails to make money in a few of these endeavors, he visits the TANU offices in search of a job. There, he is told that there are no jobs in the city and that he should instead join an ujamaa village. After ignoring this advice, bad luck befalls him in the form of being attacked by a street gang. This experience convinces Yombayomba to leave the city and ultimately return to his own village, bringing with him the principles of ujamaa. Indeed, the closing moments of the film reminds viewers that the poor man's best "stick" with which to fight off poverty was ujamaa.

A reviewer of the film wrote in the *Daily News*, "*Fimbo ya Myonge* is a must for every Tanzanian who has our Ujamaa construction at heart for not only does it depict the rich cultural heritage of our people, but also goes all out to show the ills of urban living and the benefits enjoyed by people leading an Ujamaa life in a rural environment."[84] To be short about it, this was a film about where men like Yombayomba belonged, and it was certainly not in the city. And yet for all its billing as a film about ujamaa, *Fimbo* had striking similarities to the film *Muhogo Mchungu* (Bitter cassava), produced by the Colonial Film Unit in 1952, about a "rural bumpkin" trying to make it in the city only to be "robbed beaten, hoodwinked, and hit by a car before deciding to return to a quiet and productive life in the village."[85] Not incidentally, the lead in *Muhogo Muchungu* was played by a young Rashidi Kawawa, the future prime minister.

It was not only the state that engaged in a moralizing conversation about the evils of city living through the aesthetics of storytelling. In the prolific Swahili literature and music to emerge during the socialist period, artists and intellectuals also took to the question of where Tanzanians belonged. In many of these narratives, leading a rural life was moral and responsible while the siren call of Dar es Salaam was ethically and economically ruinous, breaking families apart.[86] And like in *Fimbo*, in these stories leaving Dar was the path to redemption for both the protagonists and the nation. As men, women, and young "school leavers" found their way to Dar, rural livelihoods took on an outsized moral position in narratives like *Fimbo*. In them, urban life had the power to render young men and women unfamiliar even to themselves: the ultimate price of being in a place you did not belong.

During Operation Rwegasira, *Fimbo ya Mnyonge* toured the Dar region in a mobile film van for free showings in eight villages and across eighteen urban wards as officials hoped to garner excitement about leaving the city for the outskirts.[87] The local TANU offices also established a yearly cash prize for the best ujamaa village in the Dar es Salaam region. Geza Ulole, a village in Kigamboni, which is a brief boat ride away from Dar's harbor, won the prize in 1976 after having captured the title the previous year as well.[88] The Urafiki Textile Mill band also composed a song about Geza Ulole encouraging people to move to the new village for a better life.[89] In Zaramo, the name of the

2.7. Regional Party Secretary Rwegasira giving visiting dignitaries from Yugoslavia a tour of Geza Ulole, 1975. Photo courtesy of Tanzania Information Services.

village meant "try and see," and it reportedly grew from 217 members in 1969 to just shy of 1,200 in the intervening seven years.[90] The village had started its own fishing project and a successful poultry operation as well. "Their future plans include brick laying, charcoal and lime burning. It is hoped that these activities will boost their development both economically and socially."[91]

Despite the state's ardent efforts to inspire rural belonging through both corporeal and social compulsion, most urbanites forcibly relocated into Dar's hinterland simply waited until the peoples' militia had departed to take off on foot back home to the city, hoping in the future to avoid the "police swoop."[92] As the *Daily News* reported, many village sites where hundreds of jobless people were dropped off were virtually abandoned just days later. In one such village, when important party personnel were scheduled to come through to examine their progress, local authorities tried to disguise the recent exodus by rerouting the tour to more settled villages. In another instance, the impending visit by Rashidi Kawawa the following day, regional authorities resettled twelve families into one village in the middle of the night so there would be people to greet the prime minister.[93]

To enforce these campaigns, the state turned to Dar's institutions of urban communal belonging to now expel people. In his fervor to reorder the city Rwegasira relied on ten cell leaders to escort the peoples' militia from house to house "to identify the unemployed."[94] He also warned against trying to hide under the protective auspices of "mysterious cooperatives which have been mushrooming in Dar es Salaam since the war against loiterers was declared."[95] The very ways of finding resources and building urban communities were now being treated as suspicious havens for those who belonged on the other side of the boundary line.

Larger neighborhood authorities also became instruments of expulsion, echoing the self-policed exclusion of middle-class neighborhoods in the late colonial period. Kitabu Kata, the ward secretary of Magomeni—the neighborhood that perhaps best exemplified late colonial entitlement—made sure that anyone not gainfully employed in his ward was resettled during Operation Rwegasira. Furthermore, Kata argued, "Before a person's identity card is considered valid by the local authorities the person concerned must prove that his monthly income allows him/her to meet the Magomeni standard of living adequately," assuring that everyone in the neighborhood was part of the salariat.[96] The peoples' militia roamed the neighborhood, keeping out any small-time businesses such as "hawking groundnuts, peddling cigarettes, selling ice-water," and according to the ward's secretary, the "local court, and the police station adjacent to it" loomed large as "constant reminders to the unwanted guests of Magomeni to stay clear."[97] In this way, ten cells and neighborhoods proved a remarkably effective means of transforming the cityscape even when they were instruments of exclusion rather than conduits for development.

By the end of November, the *Daily News* reported the success of Operation Rwegasira as evidenced by "less crowds at bus stops, markets, and other places where the jobless used to congregate."[98] And yet this explanation clearly avoids another obvious reason why fewer people may have been on the street: occupying public spaces in the city had become a liability. One Dar resident, Ally Kinongo, recalled to me in conversation, "Being caught was a normal thing. If you were caught and you knew how to provide a good statement to the police it was easy not to be sent [to Geza Ulole]. For example, during that time I owned a *shamba* [farm] in Kiparang'anda [an area close to Dar and still part of the region], so if I got caught I would show my shamba identity card, and I told them that I'm just here to do business but I have a shamba and family in Kiparang'anda. . . . They trusted my statements and I never got caught again."[99] Kinongo was probably lucky to have only faced being caught once, since "operations" like Operation Rwegasira were intermittently announced until 1984, when Nyerere inaugurated the most notorious campaign to rid the city of unemployed, Nguvu Kazi, which I will return to in later chapters.

While these operations were not always successful at permanently relocating the city's "unemployed" into rural areas, it is crucial to note that urbanites were already reaching into the peri-urban zone on their own. In fact, those who wanted to find land in the periphery for settlement or cultivation had an easier task of it *because* ongoing campaigns to create ujamaa villages left cleared and cultivated land abandoned in some areas. Residents like Kinongo had likely sought a small plot of land to farm to comply with government directives for urbanites to farm: it is clear that having a shamba card offered him protection against expulsion from the city. But he also farmed to help make up for more generalized shortages that were a daily reality of urban life. Kinongo's recollection of never again being caught reveals how urbanites learned to navigate officialdom in clever ways. But it also highlights the endless negotiability of urban life that always existed side by side with these rigid operations of expulsion. Honing a set of skills for both engaging with and evading authorities was at least as important to urban survival as having the right official papers and permits.

When I interviewed Kinongo in 2009 he was living in a house he had built in Mbagala, now a densely inhabited neighborhood at the city's southern antipode that was only sparsely settled for most of the 1970s and early 1980s. In several interviews, residents routinely described Mbagala at that time as *kama kijiji* (like a village). These men and women did not necessarily settle in Mbagala in the period of these directives but many of them were generally moving toward the outskirts of the city at this point.[100] And like Kinongo, most were not arriving in the outskirts straight from rural villages but from the heart of the city, after spending years in a variety of Dar's formal and informal neighborhoods. One Mbagala resident recalled, "in Mburahati (a more central neighborhood in Dar) life started to be harder, you were required to purchase everything, but here in Mbagala you can get firewood from coconut trees and mango trees, free water from wells, so life in Mbagala was easier, even to cultivate was possible."[101]

These men and women were at different times factory workers, truck drivers, fishmongers, and furniture makers. Now they had begun to offset urban insecurity with access to cheap land where they could farm or slowly build their own homes with what resources could be found at hand in the peri-urban areas. Many who settled in Mbagala were originally from the Rufiji River region and in part ultimately chose the neighborhood because they could follow Kilwa Road south to their families and villages, walking for days if they couldn't find a ride. These new residents were not dissuaded by the lack of basic community infrastructure, even though this frequently figured into the quick decampment of expelled urbanites from regional ujamaa villages.[102] They were choosing these places for their own reasons on their own terms. With urban jobs and wages decreasing, urban families traded closer access to markets, bitumized roads, wage labor, and electricity for the relative invisibility of the margins and the advantages of going to ground somewhere between the city and country.

The area of Mbezi Beach in the northern region of the city offers another example of the layered settlement of the city's periphery through villagization, government directives, and self-settlement. In the early 1970s, Mbezi was home to only around two hundred residents, most of them workers at the nearby Tanganyika Packers meatpacking plant. The colonial state had first encouraged plant workers to settle in the region surrounding the plant because the hilly landscape was better for "hand cultivation" than "by machines favoured by the European or Indian," making it an advantageous mix of urban and rural practices for workers in the area.[103] A decade later, four thousand residents called the area home. This sparsely settled area first started to grow under Operation Pwani, when local Zaramo and Ndengereko farmers were moved into ujamaa villages.[104] Others in the area sometimes avoided villagization because they could claim seasonal work at Kibamba and Kimara brick factories. The next year, when Operation Kilimo cha Kufa na Kupona was announced, government offices and factories were also assigned land outside the city for farming "to produce at least some food for themselves."[105] Three national banks of Tanzania (the rural development, investment and housing banks) combined efforts and started a twenty-acre farm in Mbezi near the Tanganyika Packers to grow cassava, bananas, cowpeas, and pineapples. Their longer-term goal was to create a "'model farm' for the benefit of the peasants in the neighbourhood."[106] What had been sparsely populated was now layered with a palimpsest of recent incursions.

Decades later, when Mbezi residents were interviewed for Tanzania's 1994 Presidential Commission on Land Matters, they recalled being relocated to Mbezi in 1974 following state instructions to farm, only to have their land taken away just two years later as the region was declared a "planning area."[107] When the state began allocating plots for building on this new land now designated "urban" property, the original tenants were not given the opportunity to purchase this land nor compensated for their relocation. With this massive reshuffling of peri-urban space for villagization and Operation Kilimo, land frequently fell under multiple authorities: informal sale documents were kept by village *wazee* or ten cell *wajumbe*, land could be allocated as part of an ujamaa village, or by the Ministry of Lands. Contestations of ownership could be dizzying and confusing to untangle due to the omissions and overlaps of these titling systems. Opportunists became adept at exploiting both these layers and their gaps. The state had not necessarily considered the call to farm as a permanent dispensation of land to new communities but rather as an emergency measure. As the region grew, already leased land was sometimes resold again to different parties or claimed for zoning and "planned" plots by the city. Those who had been relocated explicitly to engage in rural activities and village life now found that their own development efforts had led to expanding the urban boundaries of the city and thus their own renewed alienation in the name of urban planning.

The role of the urban rent seeker is one that Nyerere had hoped to abolish through the Acquisition of Buildings Act, but landlords and renters of all races have long been common characters in Dar. During the colonial period, an impressive number of Africans owned houses and collected rent to subsidize their lives in and out of the city. This was particularly true in the peri-urban zone where Africans squatting on government land would sublet to others. For both men and women, owning a home or land and renting it out was an enduring form of financial security and one of the most important things you could obtain over the course of your life in the city. The Swahili-style house I discuss in the following chapter also allowed for families to take on boarders to cushion their weekly expenses. But beyond buoying families, by the end of the colonial period there was also intense commercial speculation driving up prices and ensuring that the most appealing land around Dar was generally moving from communal ownership into the hands of a salaried class of urbanites.[108] It seems, though, that during the 1970s as the state forced the "unemployed" into regional ujamaa villages while also requiring both families and factories to farm, this perhaps opened up a brief window of opportunity for the lower classes to obtain land quite cheaply or even freely in the urban periphery as a primary strategy for urban security. Even if these land transactions were ultimately deemed informal and thus left the owners vulnerable to eviction, in crucial years of economic crisis they could also bring immense opportunity and security to families. That window seems to have closed by the early 1980s, as salaries plummeted and middle- and upper-class urbanites also came to claim land to supplement their jobs.

Since the 1980s this has generally led to an aggregation of peri-urban land in the hands of a middle-class salariat who by then were also turning to rural regions in the city much like the rest of Dar's residents. For urban elites, hesitant to invest in building because it risked being taken away by the state, the periphery became a place to buy land, start farms, and hire laborers or families to live on them and tend them. "In so doing, they [the elites] can grow their own food to top up their meagre salaries, build weekend houses and practice farming, since most of them come from areas where agriculture is the main occupation."[109] In this way, the same salaried urbanites that frequently excluded poorer urbanites from their neighborhoods were now taking land settled and cleared by these expelled families and creating a valuable land market where it had once been virtually free.

The state, well-appointed urbanites, and the city's unemployed all turned outward as urban life became marked by food shortages as well as declining wages. Yet clearly there were still competing plans and visions of the future. For the state this was a way to solve urban shortages and repatriate farmers (in the form of unemployed or informal laborers) to the land. Yet for those who were expelled or who went on their own accord into the periphery, I would argue this was not a retreat from the city so much as a reconstitution of what city life would look like in the face of urban crisis—something perhaps new in its scale but not in its strategy for urban survival in African cities. Not only could urbanites be flexible but the city itself was an incredibly mutable environment that offered its own materials for belonging when other paths to security remained elusive.

This chapter has examined a shifting cast of state institutions and infrastructures that have attempted to adjudicate who could call Dar home and where they would live. In doing so, these institutions have fundamentally shaped the urban environment, not just within but beyond the city's official boundaries. Struggling for space in the city has remained a major preoccupation of urbanites since the colonial period. During the late colonial era, British urban planning laws demarcated both racial and class segregation in the city. Africans who had proof of formal labor enjoyed privileged access to nuclear family housing and supplemental material entitlements. But unplanned neighborhoods built by Africans rather than the colonial administration outpaced formal neighborhoods as urban migration increased after the war. These neighborhoods existed in a liminal state of administrative acquiescence but also expressed different ways of inhabiting the city.

After independence, Nyerere hoped to erase the urban scar of Dar's "slums" that had given contour to the colonial era's boundary lines of belonging. In the 1960s this hope took the form of desegregating neighborhoods, commissioning a master plan, and building more housing. But the state ran up against the reality of limited funds and rapid urban growth; they also shifted their priority to rural development after the announcement of ujamaa in 1967. Because the planning standards for housing were not changed even after the end of slum clearance, most urban residents still could not afford to meet the requirements of self-help building in planned neighborhoods.

With the dissolution of the city council in 1972, urban planning and infrastructure development as well as home building occurred predominantly through neighborhood coordination or through parastatal industries and factories. But even as these institutions became the primary tools of development they also functioned as gatekeepers of the city, mirroring the colonial legacy of urban inclusion for wage earners and exclusion for the poor. The regional government in the meantime announced a series of operations to deal with food insecurity and as part of regional villagization efforts. These campaigns expelled those without work from the city and expanded the urban periphery.

This is a narrative not of discrete instances of change but rather more ragged edges overlapping both spatially and temporally. Urban and rural spaces overlapped and could turn from one to the other and sometimes back through directives to farm on city land or the establishment of new planning zones beyond the old urban boundary. Old colonial planning standards stubbornly persisted as housing cooperatives encouraged new forms of building but were met with familiar financial barriers. Ten cell units offered a form of community building but were also tools for policing urban belonging. The following chapters will examine the practices, environments, and infrastructures that emerged in the process of drawing these sometimes overlapping and ragged boundary lines and the livelihoods these environments enabled or foreclosed.

Before shifting to these practices, I have aimed here to demonstrate how the city's "informal" expansion was not simply due to a lack of planning. Instead, this expansion was a product of institutions of planning and their colonial history. Furthermore, the expanding city reflects the intricate planning and forecasting of families and local communities attempting to navigate the vagaries of local conditions, regional "operations," and national economies by going to ground and organizing their own development efforts. The failure of planning has become convenient shorthand for evaluating African urbanism as "informal," and this too often conflates the relative material transience of these places with a lack of history. But Dar's expanding city off the map of the sewerage system, waste removal, transportation, and water was not antithetical to the middle-class city. Indeed, it was an essential part of it.

The "unplanned" spaces of colonial and postcolonial cities have long propped up the much smaller spaces of the "planned"—not just as the refuge of a surplus labor force but also by acting as an environmental middle ground where urbanites produce food and graze animals as well as excavate and sell building materials or charcoal. To consider the spatial element of urban belonging demands we consider who constitutes a city, out of what materials, and who draws its boundary lines. The very identification of these areas as peripheries rather the city itself is part of a long legacy of racial, class, and material segregation that have relegated the "African" parts of cities as marginalia in urban history. But as Catherine McNeur's book on antebellum Manhattan makes clear, it is not just cities in the Global South that are marked by "informal" economies and their accompanying heterodox environments. Manhattan was, as McNeur puts it, a city locked in a battle between the poor and rich over "what belonged in the city."[110] In the case of Manhattan, parks won out over grazing pigs long ago in a narrative that has become teleological in environmental history. But where New York was tamed, it seems Dar was, for lack of a better word, "wilded." Most urban histories narrate the process by which what is "urban" is sorted out and distilled from what constitutes the "rural." Even as the party continued to exalt the rural and vilify the urban, Dar's economy and landscape by the 1980s rendered this distinction ever harder to make.

Chapter 3

BUILDING

Materializing the Nation

In July of 1971, James Ayres, at the British High Commission in Tanzania, wrote a telegram to the Commonwealth office in London titled "Trouble at Hill."[1] The telegram described unfolding events at the University of Dar es Salaam campus involving the student activist Symonds Akivaga. Akivaga had written a letter in the radical left school newspaper critiquing the university chancellor's implementation of a disciplinary code aimed to quiet the politically engaged Dar es Salaam University Student Organization. In response, the Dar es Salaam Field Force Unit arrived on campus, and the political economist John Saul recalls, "I saw my own student . . . having been summoned for a meeting with the principal, being dragged, at gun-point, down the cement stairs at the front of the building, tossed like a sack of old clothes into a waiting army vehicle and sped away."[2] Akivaga was being "rusticated." First sent to a rural area to work, he was ultimately sent back to his home in Kenya. Akivaga was not the only student during the period to be rusticated; Nyerere and other top TANU leadership had come to view university students as privileged and out of touch with the hard work it would take to build Tanzania's rural future. Following these traumatic events, students went on strike and the administration even discussed closing the university and sending all the students to ujamaa villages to acquaint them with the realities of nation building beyond the ivory tower. At the end of Ayres's telegram laying out this troubling incident, he also noted that Nyerere "has (as so often in times of crisis) faded into the countryside—he is making bricks with the ujamaa villagers of the Dodoma Region."

My interest lies in what Saul's description and Ayres's final comment unintentionally set up as a central tension of nation building in the 1970s. Akivaga's ejection and the subsequent student strike unfolded amid the very materials of Tanzania's ongoing construction: on the one hand was concrete and on the other was bricks. Akivaga was being ejected from the well-heeled modern campus, dragged down its concrete steps while Nyerere was in Dodoma—the current site of massive ujamaa operations—making bricks

3.1. President Nyerere (right, foreground) making bricks in Dodoma, 1970.
Photo courtesy of Tanzania Information Services.

with villagers. Nyerere's activities in rural Dodoma was a clever photo op at a time of crisis: his humble productivity sharply juxtaposed the letter writing and student activism of privileged university students. While he embodied practical nation-building efforts, the students appeared disruptive and distracting.

It is unlikely that the symbolic power of these diverging landscapes and their materials was lost on Nyerere as the incident unfolded. Indeed, he had always been wary of the privileged campus on the hill he had helped plan. If any institution symbolized the fraught politics of ujamaa, it was the university. "Every time I myself come to this campus," reflected Nyerere in 1966, "I think again about our decision to build here and our decision about the types of buildings. . . . We do not build skyscrapers here so that a few very fortunate individuals can develop their own minds and then live in comfort, with intellectual stimulus making their work and their leisure interesting to themselves." For Nyerere, the concrete edifices of the university on the hill could only be justified if these "young men and women . . . become efficient servants" of Tanzania's peasantry, themselves making bricks in rural villages.[3]

Without meaning to, the Akivaga incident captured the fact that in the unfolding politics of Tanzanian development, materials mattered. This chapter explores colonial and postcolonial categories of building materials, the production and promotion of cement and bricks, and then turns to the processes of home building in Dar es Salaam. Due to expanding villagization efforts, the 1970s would become a decade of intense construction for the new socialist future. Bricks and cement were cast as the central materials of ongoing national transformation, sometimes collaborating and other times competing materials of socialism and sovereignty, also tracing a divide between city and country. Tanzanian newspapers printed stories regaling brickmaking operations or reports noting the recent output from the Wazo Hill cement plant on the outskirts of Dar. In January of 1975, for example, the *Daily News* printed a multipage story with detailed instructions for how to make your own burnt bricks.[4] But concrete and bricks had long been more than just practical construction materials. As materials of the colonial period as well, they were already imbued with politics, morality, and the seemingly predictive power to determine individual and collective futures.

By focusing on building as a set of materials and practices—turning only briefly to architecture and the professional construction industry—I hope to tack between and ultimately bring together a few different conversations. More than any other aspect of house construction or the resulting structures, building materials were imbued with narratives of science, technology, and modernity; bricks and concrete were assemblages even before they were used to build a house. This chapter traces the promotion of certain construction materials over others by the colonial and postcolonial state in order to chart a technopolitical history of modernization. But this story also traces the state's attempts by the 1970s to create a vernacular, autarkic alternative to concrete, that was nevertheless still rich with transnational intellectual and economic connections. Whether with bricks or concrete, the practice of home building (and building a new nation) was always embedded within the "imperial debris" of Tanzania's colonial past and shaped by ongoing global economic realities.[5]

Methods for producing both concrete and bricks were also just as crucially scrutinized as were the resulting materials. Planning a nation, planning a house, and planning a cement plant were deeply contingent enterprises, and the state urged citizens to be aware of the precarious relationship that held these enterprises together as well as their responsibility for sustaining that interdependency. Making bricks, on the other hand, did not so readily highlight national vulnerabilities, but instead tapped into the seemingly never-ending potential of self-help labor. In this way, building materials and their production capture much larger negotiations between citizens and the state regarding what was an "infrastructure" of the state and what would exist only through the labor of citizens.

In the final section of the chapter, I turn to how people constructed with these materials. Once conscripted for building, all materials were recast, frequently not fitting easily within national modernization narratives and instead capturing the reality of dwelling in an unknown state.

Colonial Technopolitics of Building Materials

Colonialism was perpetually wrapped up in a "fundamental structural tension between preservation and transformation" of African culture and society.[6] Cityscapes—their infrastructures and materials—were one of the central stages where this tension could be felt. As the literature on urban Africa has shown, "the African" in the city caused colonial administrations grave anxieties regarding not just law and order but also the effect urban life might have on the fabric of African communities and cultures. And yet in the "second colonial occupation" after World War II, colonial rule became deeply infused with a language of obligation to create modern subjects and consumers in work, dress, hygiene, and housing.[7] The path to modernity for Africans in these colonial narratives was, to a considerable degree, through engagements with material objects and technologies. The house, and more broadly, the infrastructure of cities, education, and transportation, were crucial tools in this emerging technopolitics.[8] Houses became structures impregnated with the ability to elevate their inhabitants and were seemingly able to do so without overtly invoking politics. One example of how these colonial narratives were dispersed is the film *The Wives of Nendi*. Produced by the British Central African Film Unit, the film follows one woman, Mai Mangwende, in her attempts to transform a village in Rhodesia from "dirt and sickness" to "tidy and healthy homes."[9] The messy homes are inhabited by the wives of a man named Nendi who are portrayed as lazy, selfish, and unkempt. When Mai prevails in reforming the village, all of Nendi's wives cheerfully sit down in dresses to a table of tea and cake they have made. Material transformation and moral conversion are simultaneously achieved and interdependent.

But within the context of cities that remained largely European enclaves, access to these material conduits of modernization were highly regulated and open to relatively few Africans. In Dar, it was not finished housing so much as building materials themselves that were imbued with qualities of modernity and tradition. Palm leaves, timber, sand, mud, bricks, and cement were not distinguished by their properties for building but rather by sorting them into categories of "traditional" (or "nonpermanent") and "permanent" materials. The colonial administration racially segregated and legally regulated neighborhoods in the city by these material categories. European and Asian neighborhoods were built in concrete and brick while most Africans—particularly new urban migrants—built with "traditional" materials in unsurveyed areas of the city. Equating "impermanent" materials with "tradition" rendered them undynamic and unadaptable, suggesting that Africans used mud, palm leaves, and wooden saplings out of rote habit rather than through consideration of their relative merits, economy, or availability. This division also missed the fact that mortared houses made with clay, lime, and coral had long been built by Africans along the Swahili coast.[10] Lastly, as I will return to later in this chapter, temporal divisions of materials as "permanent" and "impermanent" did not represent any reality of how Africans used these materials or saw their own tenure in the city. Rather, these temporal markers of materials were legally useful categories. As I have laid out in the previous chapter, the colonial state sought to offer proportionally few Africans permanent entitlement in the city through the control and construction of middle-class African neighborhoods. These were to be built out of permanent materials.[11]

Reifying houses according to their materials allowed the colonial administration to exert legal control over the quickly changing urban landscape while not bearing the financial burden of a massive extension of urban infrastructures. Areas of Dar that remained beyond the city's boundary were open to building without municipal oversight using mud, palm leaves, stones, timber, and corrugated tin. As Leslie remarks in his survey of the late colonial city, a surprising number of African landlords in Dar es Salaam (8,000) were able to erect houses "as he thinks best" and rent out rooms in houses in these areas.[12] And yet these same builders were legally prohibited from building or improving with permanent materials of cement or bricks.[13] Keeping these houses "impermanent" made it cheaper to compensate owners anytime the city claimed these regions as "planning areas."

Most of these houses were constructed in the "Swahili style."[14] Swahili houses are rectangular with a central corridor and four to six rooms off the corridor for sleeping. Their shape accommodated large and extended families or created the opportunity to rent out rooms. If there was enough money for such things, each room in the house could be outfitted with a door and a padlock, allowing privacy for renters. If all the rooms were not rented, perhaps the kitchen would be moved inside from the courtyard. In essence, they were hugely adaptable living spaces with much of the domestic life of home taking place outside and the rooms left largely for sleeping. They also were not regulated for overcrowding and sometimes as many as twenty to thirty people were recorded as living in the same house.[15] Frequently, these were incredibly ethnically diverse places of lodging as well.[16] The flexible appeal of these houses was not mirrored in the colonial built "quarters." Instead, the architecture of these buildings both assumed and attempted to instill configurations for nuclear families. Likewise, colonial officials noted that "Christian educated men with a stable marriage which he is loath to lose" preferred the "government quarters" while "uneducated and illiterate" coastal Swahili with "loosely knit marriages or less" prefer the "ready-made club-help in sickness" and companionship of Swahili houses.[17]

Creating legal distinctions of materials to demarcate the tenure of neighborhoods within the city was certainly not unique to Tanzania nor to the British. Mozambique's capital, Maputo, developed under the Portuguese as two materially distinct cities throughout the colonial period.[18] The centralized European enclave was colloquially known as "Cement City" while the surrounding "Cane City" that covered the "hills and swamps" of the peri-urban zone was home to half a million Africans until independence.[19] Their houses were made out of sugar cane reeds, asbestos cement sheets, and sheet metal. The Portuguese also prohibited building in permanent materials for Africans until there were "proper planning and services"—a perennially delayed point of departure.[20] Thus the relative autonomy in building for Africans also relegated urban communities to perpetual impermanence while providing a cheap loophole for the colonial administration to simply not provide enough housing. African housing teetered on the fault line of infrastructure in the city; these neighborhoods inhabited the liminal space between what and who would receive access to roads, water, plumbing, electricity, and what regions were merely the future city, one day deemed a planning area, cleared, and then planned for "permanence."[21]

The dearth of permanent housing for Africans during and after World War II was also due to global shortages in building materials, which drove up construction costs and sparked a flurry of research to find the magic bullet of the inexpensive house. For the British, the answer to this problem lay in turning the "craft" of homebuilding in the colonies into a science. Throughout the 1940s and early 1950s they opened up building research stations, first in West Africa, South Africa, Australia, and India, and by 1948 every British colony was assigned a technical officer to gather information on different housing and building techniques.[22] Each officer was tasked with finding ways to reduce the costs of building through experimenting with lowering building standards and identifying materials and techniques that could both be locally sourced yet scientifically tested. Where traditional craftsmanship of housing in Africa relied on what technical officers derisively considered

"rules of thumb," these new methods would instead emerge from the labora-
tory. And where traditional materials could not be used for modern houses
because their "blindfold application . . . was a gamble," technical officers applying the science of tropical building would find materials and styles that
considered "building climatology, thermal conditions, natural ventilation,
day lighting, and the use of solar energy in the tropics."[23]

Specifically, Dar's officials hoped to find materials and building methods cheap enough that the colonial state could avoid subsidizing rents for
African housing. The colony's planners sought to design a house that would
define a new building standard somewhere between what was considered
temporary and permanent. The result would be a "permanent two-roomed
quarter, except that the walls above floor level would be of green brick instead
of concrete blocks." They hoped these houses, when built by a local contrac-
tor, would cost a total of around 100 pounds, allowing the state to charge 20
shillings a month in rent to combat the rising price of rent: "When we have
constructed a few hundred of these houses, there is every reason to think that
the market rent will cease to rise. Later we hope that we shall force it down.
A social service can be an economic weapon too."[24] Anthony Atkinson, the
colonial liaison officer for East Africa's Building Research Station, was far
more skeptical of the search for a cheap solution: "Too much hope is put on a
new method of building to produce a 'cheap house.' There is no magic material which will build the perfect, permanent house for a few pounds, though
quite a few people in East Africa still seem to believe there is."[25] Cement, for
example, was three times the price it was in South Africa and double the price
in India. Likely this price difference was part of what prompted a flurry of
research into the use of alternative materials.

The German architect and city planner Ernst May is another character in
this story of the search for the cheap and efficient tropical house. Having left
Germany for East Africa in 1934, he designed a new master plan for Kampala
complete with a prototype for prefabricated African housing. May argued
that Africans should hone their own "'typical African style of [modern] architecture,' and to do so on an 'economic basis' in a manner that they could
afford."[26] May made his own prototype from clay shingles and prefabricated
concrete arches, appearing not unlike a concrete Quonset hut. May believed
Africans should eschew European designs and instead live in something that
spoke to the "psychology of advanced natives." It is ironic, then, that he was
offering up for the cause his own design, which bore striking resemblance to
work he had done twenty years prior in northern Germany. His designs never
took off because Ugandans instead wanted the same houses that were being
built for Europeans.[27]

With the end of colonial rule, anxieties over independence among colonial administrators and white settlers frequently manifested through con-

cern for the continent's material infrastructure if left in the hands of Africans. European administrators imagined the end of empire as a traumatic disruption of progress where the material legacy of European rule would simply decay. Who would maintain and care for these buildings, making sure they did not disintegrate? Perhaps the most extreme example of this rhetoric emerged with the end of Belgian rule in the Congo. In a 1960 *Life* magazine article titled "End of Era with Threat of the Jungle Taking Over," a reporter asked the governor-general of the Congo what would happen following independence: "He replied that grass would grow in the streets. The buildings may last 10 years . . . because we have built well, but after that the jungle will close in; it will take over what we have wrested from it." The journalist goes on to note that in the Congo, "mold grows on shoes overnight. Only unremitting supervision can arrest these elemental forces, can hold them back in the endless war against the terrible fecundity of the forest."[28] Indeed, the very justification of colonial rule was that it offered the "unremitting supervision" needed to navigate the ongoing tensions of preservation and transformation. The threat of decay or infrastructural recidivism acted as justification (and later cause for nostalgia) for a continued colonial presence.

Ann Stoler, however, turns those imperial fears on their ear: it was not the anxiety of a biological recidivism that lurked for new nations but the reality that new nations were always building among "imperial debris." Dismissing the sometimes vague notion of struggling with a colonial legacy, Stoler argues that these legacies were never so ineffable. New nations were instead stuck with emerging from the physical presence of the empire's "ruination." This framing highlights how new nations were literally occupying the remains of the old regime.[29] Thinking with ruination highlights how processes were still unfolding from these old "imperial formations."[30] Nyerere himself offered a similar insight connecting the material and spiritual project of emancipation facing the new nation: "A man who has inherited a tumbledown cottage has to live in even worse conditions while he is rebuilding it and making a decent house for himself. In the same way we have to accept the existence of problems which are created by the very fact of trying to convert the colonialist and semi-capitalist economy we inherited into a nationalist and socialist economy."[31] In effect, new building was always happening within a larger context of debris from the past: while colonial structures might not have all stood the test of time, the colonial categories for materials and their fraught valences of modernity and tradition certainly did.

After independence, Nyerere did not rebuke colonial categories of permanent and impermanent nor untether them from their connotations of modernity and tradition. Rather, he sought to strip the categories of their racial and colonial connotations and integrate them into his own evolving governmentality.[32] The new nation's early policies of urban slum eradication acted as a referendum on the colonial relegation of Africans to poor living conditions and yet also offered continuity with the colonial administration's attempt to intervene against "backwardness." For example, the National Housing Commission's slum clearance program was couched in medicalizing terms similar to colonial sanitation narratives blaming the materials of makuti houses (a house with a palm thatched roof) for the spread of diseases such as tuberculosis, measles, dysentery, and malaria.[33] Also like the British before, the state saw the use of permanent materials as a didactic tool of modernization. In addition to converting Tanzanians to users of concrete and bricks, policymakers encouraged builders to become "modern consumers of wood."[34] Instead of gathering and cutting small poles, saplings, and fuelwood, home builders were instead to seek out sawed timber and plywood. This transition would aid in ending "the impermanent shifting life that is the root cause of poverty in the midst of plenty."[35] Converting Tanzanians to better building materials was thus part of state education campaigns rather than just calculations of cost or supply. When announcing that 1970 would be the year for adult education, Nyerere wrote that the "first job of adult education will therefore be to make us reject bad houses, bad *jembe* [plows], and preventable diseases; it will make us recognize that we ourselves have the ability to obtain better houses, better tools and better health."[36] In a rhetoric echoing colonial campaigns, Nyerere saw the house as the container by which a reluctant peasantry would be pulled out of the past and remade into ideal citizens.[37]

Despite this supposed reluctance to change, in the years leading up to the adult education campaign, Tanzanians had in fact displayed an eagerness to build with permanent materials. Cement in particular provided security for those who could afford it. Building homes was not some intractable preference for "traditional materials" but involved questions of experience, aspiration, availability, and money. If anything, communities seemed all too eager to build with cement. For example, when dispensing cement donated by UNICEF, development officers were warned to keep it under lock and key so it could only be used for community projects.[38] Michaela von Freyhold, in her case study of the forced settlement of Segeda village, also commented that the construction of the village's community center was the only place where men "worked with enthusiasm and without coercion."[39] After the completion of the project von Freyhold noticed that their success was "having its psychological effect on the villagers":

> The first effect stemmed from the stark contrast between the community centre and the dwelling houses. A very "concrete" utopia was taking hold of the imagination of the builders: good modern houses for the villagers to live in. Since the building materials for the community centre were provided by the government,

the villagers hoped that they might get government aid for their houses as well. This suggestion was turned down and people were told to wait until they could finance such an undertaking themselves. The success of the building brigade demonstrated the satisfaction which peasants derive from the utilization of more modern methods of production.[40]

Perhaps generations of colonial and postcolonial rhetoric had worked? Or perhaps villagers also shared the sentiment that permanent materials were indeed better for both building and living. But inextricable from that desire for more modern houses was that permanent materials could offer families a bulwark against state resettlement. While urban slum clearance was no longer policy after 1972, as forced villagization unfolded in the next year across Tanzania millions faced the task of relocating and rebuilding homes. As Elizabeth Daley points out, "the destruction of houses and use of lorries to move people were common . . . while people across the region often found themselves sleeping outside at the coldest time of the year."[41] One man told her that when he was reluctant to leave his home, it was simply burned down. In their new villages many inhabited the skeletons of half-built homes as they settled in, with permanent materials in short supply. But those who had originally lived in "permanent houses along the road" frequently did not face relocation.[42] Considering the potentially immense and obvious advantages to having built a home with permanent materials, the state's portrayal of citizens suffering from an intractable preference for "bad houses" and in need of adult education deserves a skeptical reading. Cement, for its quick construction and designation as "permanent," offered multiple forms of security.

Cementing the Nation

With the shift to compulsory villagization and a range of infrastructural projects, Tanzania was in dire need of building materials by the 1970s. Some of these projects included the construction of the international airport, the Kidatu hydroelectric plant, the TAZARA railway, the extension of the harbor, and the relocation of the capital to Dodoma. Architectural historian Adrian Forty writes, "Whenever and wherever urgent modernization has been called for, concrete has been pressed into service."[43] And Tanzania was no exception: Nyerere deemed the establishment of a cement plant in Dar es Salaam an early priority for the new nation not just for infrastructure projects but as the key material for developing both a new "high" and "low" Tanzanian style of architecture. Dar in the 1960s had become the site for numerous new building projects, nearly all out of concrete and many of them contracted to house vital new state institutions. A small group of architects had begun to articulate a modernist architectural style for the new nation in this series of buildings. The city's first architectural firm was purportedly established in

3.2. Kariakoo Market, Dar es Salaam, designed by architect Beda J. Amuli. Wikimedia Commons, https://commons.wikimedia.org/wiki/File:Karia koo_view2.png.

the wake of the failed Groundnut Scheme in 1949, when the architect C. A. Bransgrove remained behind in Dar after its dissolution. Bransgrove's firm, along with the British firm Norman and Dawbarn, who designed Dar's university campus, were early designers in the city's evolving modernist style, soon to be joined by Tanzanian-born architects Anthony Almeida and B. J. Amuli. Almeida, of Goan heritage, and Amuli, an African trained in Tel Aviv, also brought transnational modernist style sensibilities to their projects such as the Kilimanjaro Hotel, Ushirika Cooperative Building, the Central Library, and the East African Community Regional Headquarters.[44]

In conversation with this first generation of Tanzanian architects Nyerere urged these men to create a national architectural style, citing China and Japan as examples. Architect Joe Noronha commented that one of the major hurdles in creating a high vernacular style for the new nation was the sense that "our culture is a village culture," but using mangrove poles and mud "to create a national identity is very difficult."[45] Perhaps the best example of this experimentation with new materials mixed with a "village culture" is the massive concrete edifice of Kariakoo Market, designed by Amuli in 1974. Amuli had designed the new market in the center of Dar's commercial area to emulate a traditional African market unfolding under the shaded branches of a tree. The resulting building looks like a series of inverted concrete umbrellas designed also to keep cool and collect rainwater for reuse.[46] Cement was quickly becoming an essential tool in creating an emerging Tanzanian architectural vernacular as well as a necessary material for nation building.

3.3. Wazo Hill Cement Plant, 1967. Photo courtesy of Tanzania Information Services.

This new demand for cement manifested its own material politics that extended both long before and long after concrete was bagged and sold.[47] Because it is bulky and expensive to ship, cement was routinely one of the first new industries slated for import substitution in new African nations. While cement can usually be produced from local materials, massive cement plants are more complex assemblages. And whereas "impermanent materials" for building are reliably found in many locales, the resources needed for concrete are site-specific. It also requires considerable expertise and capital investment.

Anticipating the importance of concrete in particular for the new nation, Nyerere had essentially demanded the construction of a cement plant after independence. Tanganyika Portland Cement Company was formed in 1959 in association with the Swiss company Cementia and the British Portland Cement Manufacturers (PCM). PCM already owned a plant in Mombasa and resisted opening one in Tanzania, claiming that it was unnecessary and against the spirit of East African Common Services Organization. Nyerere, however, remained adamant that cement was a crucial commodity for national production and responded by criticizing the unbalanced distribution of industrial and commercial activities across East Africa, which he saw as replicating the colonial habit of favoring Kenya.[48] PCM ultimately relented in order to preserve their own monopoly in the region.[49] Under the initial

3.4. Visiting government dignitaries from Guinea watching a bag of cement coming down the conveyor belt at Wazo Hill, 1970. Photo courtesy of Tanzania Information Services.

arrangement, Cementia Holding maintained majority control of the plant, with the state having a 10 percent holding and consequently excluded from much of the plant's decision making until it was nationalized in 1974.[50]

The Wazo Hill cement plant was built twenty-four kilometers outside of Dar es Salaam in 1964 with cement production beginning in 1966. Wazo Hill was, geologically speaking, a great location because of its ready supply of limestone chalk, seashells, clay, and shale. Since it rested on a coral limestone bed with another quarry for red clay a kilometer away, the factory was assured at least a twenty-five-year supply of raw materials.[51] Limestone was blasted out and then loaded mechanically onto trucks and driven to the plant. Once the limestone and red clay were crushed together and thoroughly mixed, the next step was to put the mixture into a kiln until it reached 1,400 degrees Celsius. The resulting product is known as clinker. Wazo Hill's kiln was fueled by furnace oil but others can also run on coal or gas. After the kiln, gypsum was added to the clinker to make Portland cement.

In addition to the raw materials on site, the process of producing cement required a large amount of both local and foreign inputs.[52] Beyond the imported fuel for the kiln, the cement factory relied on electrical power for several stages of production as well as to power its workshops, canteen, hospital, and housing estate. Oil-based thermal generation that relied on foreign oil imports made up almost half of Tanzania's energy production by 1973.[53] As a result, following the first oil shock, the average cost of fuel and power per ton of cement went up fourfold.[54] After the cement was produced, it was stored in silos and cooled, moved by bucket elevators and pneumatic conveyors. It was then packed by an automatic packing machine into bags for sale or for purchase in bulk. Bags were then distributed to regional trading companies and each national region received a quota amount.

And if geology, electricity, and water (Wazo Hill required about two million gallons of water a month for production) were crucial contingencies of the cement assemblage, so too were spirits, labor, and the new art of reading these machines. After an onerous and delayed construction process, the plant's expatriate general manager, wary of keeping everything online without unnecessary stoppages and repairs, asked the Ministry for Industry for advice on how to deal with inexperienced workers. In his estimation, human error would be the root cause of delays. The minister, however, instructed him to bring in "witchdoctors to exterminate the bad spirits out of the plant."[55] Nyerere affirmed the minister's idea, as did the Workers Union of Tanzania. Zvonko Springer, a Yugoslavian structural engineer hired to oversee the opening of the plant, remembered that all the male workers were brought together to slaughter two oxen under the supervision of a local healer. Each machine in the plant was then smeared with oxen blood:

> The patch size apparently depended upon the importance of a particular building as considerate [sic] by the Chief witchdoctor. We could not envisage why the largest patch was not on the Heat exchanger building—certainly it was the most important and largest one in the factory. . . . A few days after this eventful ritual was done something unbelievable happened. All the plant machinery was running almost smoothly. The green light on top of the Heat exchanger tower shone green for days without turning to yellow or even red. It seemed like the "bad spirits" have been appeased and that the plant future was the most promising one.[56]

The entire project of constructing the plant and beginning operations was a massive endeavor, intimately connected to the soil and chemical make-up of Wazo Hill, the importance of furnace fuel, the production of power, the availability of water, and perhaps most of all, the treacherous relationship between man and machine, mediated by blood, spirits, and, of course, foreign expertise.

Despite the early green light hovering above the heat exchange, the plant frequently went offline for a variety of reasons including lack of spare parts, stalled deliveries from Dar's congested port, and human error. Outages and stoppages, frequent by the mid-1970s, were particularly expensive due to the thermodynamics of cement production. Any time there was a stoppage in the clinker process it took immense amounts of fuel to reheat the kiln. Losing productivity meant fewer bags of cement to load onto the waiting line of trucks outside and the growing need to seek out alternatives. Those alternatives were either the expensive importation of foreign cement or turning to alternative construction materials and methods when possible. To the frustration of many, establishing a cement plant had been a bid for resource independence and yet its continued malfunctioning revealed the nation's vulnerable dependencies on foreign finance, expertise, and, particularly after 1973, the price of foreign oil. These dependencies, cement's expense, and its aspirational modernizing qualities could not be uncoupled from the product itself. Indeed, one of the Swahili terms for cement is *ulongo ulaya*, which translates as "European soil."[57]

As supplies waned in the 1970s, cement distribution became severely rationed for private homebuilders, giving priority to national infrastructural projects. This further marked cement as a privileged material and one connected to crucial state infrastructures but not accessible for self-help building. That infrastructure projects were waiting on the very material of their construction also became a symbolic explanation of the urgency of such projects. Stoppages pointed out the recursive and interdependent nature of infrastructure and its production. The cement plant could only work with electricity. And likewise, cement was required to expand the nation's hydroelectric projects so that Tanzania could sever its dependence on foreign energy. In turn, this required proper transportation infrastructure, because without functioning railways and roads, bags or trucks of cement would not get very far outside of Dar es Salaam to build these projects. And bitumen roads were themselves literally constructed out of a semisolid form of petroleum. Nyerere had remarked at the inauguration of the Hale hydroelectric project in 1965, "Schemes such as this one are in fact the bricks and mortar evidence of the revolution which our country is deliberately and purposely undergoing. It represents the application of science to the needs of the people."[58] If the cement plant could not produce, then other infrastructural projects—the scaffolding to independence and autonomy—were mutually imperiled.

Production levels soon became a quotidian concern of not just plant managers but the newspaper-reading public as well, remaking Wazo Hill's stoppages into the stuff of rumor and accusations of foul play. The police were even dispatched to the plant in 1972 to investigate whether the machines were being sabotaged due to continued breakdowns and chronic shortages. Initial rumors of sabotage accompanied gossip that the expatriate engineers at the cement plant had all "tendered their letters of resignation" while the chief engineer apparently went on vacation right after two machines had gone offline, paralyzing production until his return.[59] These rumors reflected and fueled fears of expatriate departure, since no local engineers were equipped to maintain and run the plant. The newspaper followed up the next day to assure readers that no such conspiracy of departure was brewing among foreign engineers. And yet another conspiracy emerged the following month: that workers were "putting iron rods and stones in the grinding machines" and "mobilizing other workers in order to retard the development of the factory."[60] These accusations brought the minister for commerce and industries to the factory to announce that he would not be happy until every corner of Tanzania had enough cement. But no refutation of the rumors was offered.

Frustrated by having so little control over the plant's management decisions, Nyerere abruptly nationalized it in 1974, two years ahead of schedule. At this time, more than ten years after independence, Wazo Hill had only one Tanzanian sales manager at the management level and not a single local technician, engineer, electrician, or chemist. The government recruited Indian management until 1981 to replace the departing European staff, and by 1978, in an another attempt to increase output of cement nationally, the government announced the opening of a new plant in Tanga, constructed with a loan from Denmark.[61] Around fifty Tanzanians traveled to India at the expense of the state to study cement technology, with a smaller group continuing on to Denmark for specialized training.[62] Despite these attempts to change the situation, by 1981, with the national economy chronically short on foreign exchange, the factory was running at 20–30 percent capacity and management was handed over again to a Swedish firm.[63] The official price of a bag of cement by 1982 had increased by 90 percent over the past five years, while the price of building materials more broadly went up by 185 percent.[64]

For most Tanzanians, cement production remained a process that failed or succeeded based on factors seemingly beyond their control: a machine could need a spare part, an engineer could get sick, the price of oil could rise, a truck could break down. The endeavor was highly local and yet globally contingent, taking an immense amount of both regional and foreign inputs.[65] As an assemblage constantly teetering on the edge of stoppage, this precarity is perhaps what made the process so prone to conspiratorial rumor and fears of supernatural intervention.[66] From the state's perspective, cement remained a vital product to nationalize, as it was the stuff of fundamental nation-building efforts. But it was also increasingly framed as appropriate only for certain infrastructure projects. The plant's location in Dar confirmed its elite status while also reinforcing the rhetoric of the city as privileged and exceptional. Rather than equalizing development across space, cement remained an urban building material.

Wazo Hill's shortages continued to punctuate the 1970s and as concrete faltered, homebuilders urgently needed a new alternative if they were to avoid mud and *makuti*. Although Tanzanians had been compelled legally through planning and didactically through education campaigns to abandon "traditional" materials, the state now encouraged builders to also abandon their dreams of a "concrete utopia." Research institutions at the University of Dar es Salaam developed and promoted burnt bricks, with the larger task of developing a technologically mediated vernacular architecture in "permanent" materials that could minimize the use of cement.

The paucity of cement and the resulting push for a new brick vernacular highlights the two general paths of technology that Tanzania and other new nations were constantly straddling. One path was the adoption and fostering of site-specific technology that required highly trained specialists to build, maintain, and repair, like a cement plant. These sites of production were domestically and internationally appraised as prerequisites of independence and modernity. The second path was to promote more mobile, small-scale forms of "appropriate technology," which were seen as emancipatory both for the individual and the nation.[67] These technologies tended to rely on local materials and inputs. For TANU, these two narratives were not contradictory and indeed existed in necessary complement with one another, particularly as larger infrastructural projects faltered. But for citizens who, because of shortages, were compelled to adopt "appropriate technology," they had very different implications regarding their own labor and time. The shifting grounds of their material future also required swiftly recoding cement as no longer aspirational and a source of security but greedy, antisocial, and neocolonial.

While the state campaign to develop and promote alternatives began in the early 1970s, as I will chronicle below, it was in 1977, when celebrating and reflecting on ten years of ujamaa, that Nyerere best articulated his frustrations with individual builders desiring cement. Since independence, the pedagogical mission of the state with regard to building had nearly inverted itself. His original mission to intervene in an uneducated compulsion toward traditional materials was now a mission to help citizens kick their unhealthy obsession with cement:

> Although we know that most of our people cannot afford the mortgage or rental costs of the cement house, we persist in promoting its construction. . . . [F]or most people the only effective choice is between an improved and an unimproved traditional house—they cannot afford the cement house. . . . The present widespread addiction to cement and tin roofs is a kind of mental paralysis. A *bati* [corrugated iron] roof is nothing compared with one of tiles. But those afflicted with this mental attitude will not agree. Cement is basically "earth" but it is "European soil." Therefore people refuse to build a house of burnt bricks and tiles; they insist on waiting for a tin roof and "European soil." If we want to progress more rapidly in the future we must overcome at least some of these mental blocks![68]

Nyerere was now urging Tanzanians to uncouple their notion of progress from an attachment to European materials. He also declared in the same speech that the nation as a whole had also been caught up in this same mental paralysis: Tanzania had fallen "into the trap of being 'modern' at all costs."

He expressed regret over the nation's investment in "large capital-intensive factories when a number of small labour-intensive plants would have given the same service . . . with less use of external technical expertise." Forced by circumstance as well as shifting ideology, by the late 1970s Nyerere was reevaluating the notions of progress he had embraced in the first decade of independence. In urging citizens to reconsider the models of development they turned to, he was also urging them to reconsider the materials they conscripted into realizing these particular futures. But when it came to curbing the popularity of cement among individual urban builders, there is little evidence that this speech was effective. Bags of cement remained a coveted commodity. It is more likely that urbanites found Nyerere's focus on "European soil" and tin roofs irksome, with the limited choices they had to secure their own futures in the city. Nyerere's moralizing about the material aspirations of Tanzanians would become increasingly frustrating by the 1980s, as families found themselves living among a constellation of shortages yet constantly exhorted to use less.

While disabusing citizens of their preference for concrete, Nyerere also hoped to change the habits of the national building industry responsible for ongoing infrastructural projects. With the economy stagnating, the only hope for sustaining growth in the building industry lay in developing local material alternatives. Without this, the formal construction industry would remain almost completely dependent "on imported materials, plants and equipment; as well as on imported management, professional, supervisory and skilled labour."[69] Instead of acting as a productive force of development, material shortages would mean that the building boom simply ate away at an already dwindling foreign exchange budget. Tanzania's construction industry was modeled after European and American institutions, which generally observed a "complete divorce of the responsibility for the design of a building, from the responsibility for its production."[70] With foreign firms directing many development projects and not taking material constraints into account, the national construction company, Mwananchi Engineering and Construction Company (MECCO), suffered major losses.[71] In the early 1970s the state took over both design and construction firms, entertaining a proposal to merge the two into a single parastatal whose objective would be "not simply profitability, but the eventual attainment of more economic building through the development and use of appropriate, simplified, standard designs" that focused on local materials.[72] Unfortunately, this new collaboration never happened.

Making Bricks

I want to return for a moment to the beginning of this chapter and the image of Nyerere making bricks in the wake of Akivaga's violent removal from the university. If concrete was not only an urban material but also increasingly framed as a "foreign" one, Nyerere's brickmaking in rural Dodoma comes more clearly into focus as an act of both political theater and pedagogy. Bricks would build the new capital region while Dar wrestled with its concrete demons. The action embodied not just the material of a particularly Tanzanian future (bricks) but also the self-help labor involved in its production. Whereas cement production was unpredictable and draining foreign exchange, the state imagined bricks as not just materially but also socially productive.

The postcolonial push to develop an alternative to cement began institutionally with the establishment of the University of Dar es Salaam's Building Research Unit (BRU) in 1972. The unit's task was to assist home builders and encourage alternatives to cement through conducting research on building construction and material technologies. This research was then disseminated through campaigns such as Opresheni Nyumba Bora (Operation Better House), launched in 1974. Echoing colonial attempts to create a vernacular architecture thirty years earlier, the campaign's goal was to promote building with materials "that do not get easily destroyed by rain and sun; built in such a way that insects will not easily enter into the house; and are able to provide a healthy environment as well as allow sufficient light and air to enter, it must have a toilet a clean kitchen and a compound."[73] While the BRU did experiment with some other new technologies, the main alternative they developed and encouraged was burnt bricks.[74]

Beyond home building, prime minister Rashidi Kawawa also ordered all public buildings in 1973 to be constructed out of burnt bricks and tiles "in places where these materials can easily be made."[75] Nyerere further ordered that every district should have its own factory for bricks and "no public institutions shall be allowed to use cement blocks unless they obtain special permission from the principal secretary of a parent ministry." To add to bureaucratic procedure as deterrent, any ministry that wanted to use cement would have to first "write a letter to the Prime Minister to ask permission."[76]

Symbolizing the ascendance of bricks in Tanzania's new built environment, Dodoma's master plan also specified the construction of its own brick- and tile-manufacturing complex, and that "every effort should be made to use local materials and the simple techniques for which we already have Tanzanian *fundi* [technicians] and expertise."[77] The *Daily News* announced that "it will be of great historical importance to see our capital—Dodoma— built of burnt clay bricks."[78] Given the difficulties and dependencies of cement production, bricks offered a new sovereign style of building. Yet as plans took shape for the new capital, Dodoma's old brick factory had fallen into disrepair, as had the factories in Arusha and Dar es Salaam. All three relied on tired Italian extruders that were suffering from disrepair, leaving machine-made bricks available only from Soni, Morogoro, and Iringa factories.[79] Additionally, Dodoma's new proposed state-of-the-art brickmaking facility, which would purportedly be the largest in Africa, was not slated to open until 1982.[80]

While promoted and researched as an autarkic project to embrace locally available materials, brick factories had their own foreign inputs, some material and some ideological. Unlike concrete, however, the foreign inputs necessary for bricks drew Tanzania closer to other socialist nations rather than the technological transfer from First World nations that fueled the cement plant. Tanzania was one of many Third World and socialist countries seeking out cheaper building materials and looking to cultivate a material aesthetics and politics around these alternatives. To borrow from Caroline Humphrey's work on Soviet housing, there was "a definite pronounced intention of the state to make use of the materiality of dwelling to produce new social forms and moral values."[81] Christina Schwenkel's work on Vietnam also serves as a particularly relevant example of a similar conjunction of materiality, aesthetics, and socialism around bricks. Schwenkel writes about collaborative rebuilding efforts in late 1970s Vietnam undertaken by East German engineers and Vietnamese citizens. Embedded within this socialist collaboration was a sense that manufacturing and building with bricks was synonymous with recovery and nation (re)building. Schwenkel argues that it is impossible not to see that bricks in particular "mattered," quite literally. The Tanzanian state also cultivated bricks as "utopic objects of desire" that "gave shape to an engaged politics of hope and belief in future betterment."[82]

Even the BRU's pamphlet "Better Burnt Bricks" (which was abridged and republished in the *Daily News*) portrayed bricks as objects of national belonging that also tied Tanzanians to broader socialist solidarities. In promoting brickmaking, the BRU encouraged Tanzanians to join the likes of Yugoslavia, Bulgaria, India, and Pakistan in staking out a different course for development than capitalist countries.[83] In a 1978 article in the *Daily News*, Lawrence Kilimwiko also offers up bricks as a broader corrective to the growing problems of Third World slums, announcing the opening of a brick factory in Pugu near the industrial corridor of Dar es Salaam.[84] The factory was built with aid from Bulgaria's Technoexport Corporation and would open upon the return of seven factory employees from a training program in Bulgaria.[85] By 1985 there were also plans for the North Korean government to fund a brick factory in Arusha. The plant, however, five hundred miles inland, would rely on oil at a time when, "for significant periods of the year, Arusha region as a whole is without any fuel oil."[86] Managers hoped it would run at 70 percent capacity. Brick production thus articulated a different set of international relationships than cement did, calling on foreign relationships with mostly socialist countries for training and education. While the state encouraged using bricks as a valorization of local resources, they also saw it

3.5. Children helping out with making burnt bricks. Photo courtesy of Tanzania Information Services.

as emancipatory in a global context, particularly in the aftermath of the oil crisis.

Despite their symbolic divergence, brick plants, like the cement plant, also ran on oil or coal, if modified, leaving them vulnerable to similar costly interruptions.[87] And for all their potential as "utopic objects of desire," factory-produced bricks were simply not as popular with consumers as cement. In part, consumers were skeptical of inconsistencies in bricks and tiles. People were disappointed by the varying quality of the ceramic tiles emerging from the factory in Mbeya because the integrity and size of the roof tiles changed as laborers got tired and applied less pressure to the machine used to stamp them out and compress them.[88] While the operations of the cement plant may have waxed and waned based on the availability of expert knowledge or the occurrence of human error, it nevertheless produced what was deemed a more uniform and reliable product. And as the complaints from Mbeya suggested, uniformity was aesthetically desired by home builders.

Beyond the push to establish regional factories and to transnationally train workers to participate in a technology framed as socialist and Third World, the opportunity of bricks—and their real socialist potential in Tanzania—lay in the promise of community rather than factory production. As written in *Daily News*, the "technology required is simple and could easily be grasped by our local people."[89] If one wanted to make bricks, they did not need a regional factory but rather they could follow the directions outlined in the newspaper.[90] Communities were encouraged to build their own kiln, collect mud and clay, and harvest wood to burn. Brickmaking fit into an ideal type of self-help building promoted by the state: it took virtually no foreign input, improved the housing stock of communities, and relied explicitly on forms of communal labor.

In practice, brick production was physically rigorous, as well as deeply intensive in time, labor, and environmental resources. Making bricks requires an immense amount of clay and soil to be dug and moved by hand, bucket, or wheelbarrow, and it also requires fuel wood for finishing the bricks in kilns. And if bricks could be made nearly anywhere—which was their best quality—it was also a process that swiftly decimated local resources, particularly wood fuel. Tanzania's director of forestry noted with alarm in 1981 that trips for collecting fuelwood in East Africa were getting increasingly longer, with round trips of more than a hundred kilometers not uncommon as forests receded.[91] During villagization, official forest boundaries were frequently overlooked and peasants encouraged to harvest on reserve land, diminishing Tanzania's forests.[92] Part of this new reliance on forests were for bricks: according to one study, the author estimated that it took 70 trees to make 25,000 bricks.[93] He further estimated that each house would require the production of 24,000 bricks, with an additional allowance for the fact that on average 4 percent of bricks broke in a burn. Extrapolating to a community of 350 homes, building with bricks would require 24,500 trees.[94] The state estimated a smaller number of 15,000 bricks for each model of their low-cost housing designs produced by the National Housing Corporation. And measured in time rather than trees, they estimated that a team of four would take a month to produce those 15,000 bricks by hand.[95] Taking rainy season into account, when brick production could pose a challenge, the same team of four could only make enough bricks in one year to build eight to ten houses.

In the case of India, where brickmaking was also encouraged as vitally important to village development in the 1950s, Daniel Immerwahr's work captures what these "small" and "local" operations look like when taken up on a massive scale. By 1953 the northern district of Etawah in India was dotted with 520 cooperatively run brick kilns employing 42,000 workers. Not calculating in trees burned or hours spent, Indian and foreign development experts saw this massive district undertaking so favorably because "most of the labor and materials for the construction were donated by the villagers themselves, and hardly any new technology had been used." "The achievement," writes Immerwahr, "was seen as not material but sociological."[96] A similar framing of brick production happened in Tanzania. The triumph of brickmaking was the communal labor itself, along with the actual bricks. The burden of forestalled cement production or unrepaired brick factories fell on the shoulders of all Tanzanians. With anxieties about producing materials with too much reliance on "foreign inputs," time—in the form of "self-help" labor—became the most domestic of all inputs.[97] In state narratives, not only would this labor produce bricks, it would also produce African socialism.

It is hard to determine how effective the call to community brick manufacture was. What numbers I did find suggest it never took off as hoped. In 1981, twenty out of twenty-five regions were participating in local brickmaking, with a total of 152,479 households producing burnt bricks.[98] The Dar es Salaam region, along with the coast region (within the orbit of Wazo Hill), had the lowest number, with only 1,500 households producing bricks, whereas in Iringa 25,000 households participated. But in all, only about 4.9 percent of Tanzanian households were making bricks.[99] It seems that communal brickmaking never became the ascendant local technology the state had hoped for. And indeed, there were several constraints that may have played a part, not least access to wood in Tanzania's more arid regions and its labor-intensive nature, dependent on seasons. But also, in urging Tanzanians that brickmaking would conscript them into meaningful socialist and developing world solidarities, the state glossed over potentially more deeply felt and remembered historical connections. Brickmaking in colonial Tanzania seems to have mostly happened at mission sites. While the labor was not likely purely compulsory, men, women, and children were frequently involved in digging, forming, and burning bricks to build schools and community churches at mission sites.[100] In attempting to create an affective socialism through the built environment, the state's attempts to frame brickmaking as emancipatory was perhaps competing with a different embodied colonial history of working within an oppressive colonial system of obligation. While Nyerere saw the demand for cement in home building as a frustrating colonial mentality, the labor tied to bricks was also part of the "imperial formation" on which the Tanzanian state was now building.

Building as Process and Social Construction

The search for vernacular building materials has a thick genealogy, and its colonial roots strangely mirrored Nyerere's postcolonial aspirations. There was tremendous symbolism wrapped up in the technologies conscripted for producing both concrete and bricks. Economies, environments, and labor, stretching from the local to the global, shaped (and in turn were shaped) by these materials. Indeed, in comparing postcolonial and colonial attempts at creating and promoting new building materials there are many continuities. Creating a material and design that was particularly "African" has always been central to the task, whether articulated in notions of racial difference or in a postcolonial language of national sovereignty and reclaiming "tradition." In both contexts, the task of creating scientifically mediated building material and design advanced very similar goals: these houses should be appropriate to East Africa's climate and environment while also remaining inexpensive and durable and incorporating locally available materials.

But, as any builder knows, materials do not on their own act out their technological promise. As thinking about materials as assemblages makes clear, bricks and concrete (and their promises of permanence) are always mediated by the environments they then become part of and how they are used. In the process of building, materials are refashioned and negotiated not just once but with every need for repair or renovation. Tanzanians who had long been building and taking care of their own homes knew this intimately. One citizen wrote into the *Daily News* regarding the National Housing Corporation's insistence on permanent materials to note that cement did not necessarily result in more "permanent" homes. "It is true that the housing condition nowadays is much better than before the role of National Housing Corporation," the letter politely begins:

> when we, the common people were living in small thatched muddy houses known in Kiswahili "Mbavu za Mbwa" [the ribs of a dog]. The motto of the NHC was to remove all these unhygienic houses and replace them with modern ones. But it is surprising to see that the houses said to be modern are really just like the Mbavu za Mbwa type, the difference being that the said modern houses are made of cement. After six to seven months of construction, you will feel pity [for] the way the modern house looks. The walls are full of cracks, the floor full of holes, the cess-pit broken and the dirty smelling waters pouring out. One has to hold his breath when in the latrine or in a kitchen. Should the NHC spokesman need a list of badly constructed houses, I am ready to give him one.[101]

Building materials were just one aspect of a much broader process shaping the landscape of cities. Skill, experience, and availability were all important mediations that could render cement less hospitable and just as impermanent as mud and thatch. Also, for all its supposed uniformity as a factory product, cement remains an "unpredictable" and "erratic" material as it ages, despite its monolithic reputation.[102] Mud and thatch, on the other hand, could be routinely and easily repaired. One could still find makuti houses in Dar in the 1970s that had been carefully maintained for fifty or sixty years.[103] Thus, even the NHC could not offer any promises of material "modernity" or permanence.

By the 1980s only a quarter of the houses in urban areas across Tanzania were publicly owned (and if extended to include all of Tanzania, that amount was 1 percent of all housing).[104] The NHC could not keep up with growing urban need. The institutions designed to guide and professionalize home building through the proper channels of purchasing surveyed plots and procuring building loans were also too costly and chronically short of the managers, engineers, and surveyors required to keep up with housing needs. These public institutions also faced their own constraints of access to building materials. For example, the NHC by law had to import materials from China and could look elsewhere for building materials only if China could not supply them.[105] Professional building was never the target of the state's push for permanent materials because it simply wasn't how people built.

As the state's repeated attempts at building education makes clear, private home builders, communities, and cooperatives were who would ultimately wrestle with the materiality of bricks and concrete, mud and thatch. Attempts to pass on skills to home builders took the form of building brigades in rural areas and the dispersal of educational materials in both urban and rural sectors. Sometimes these educational endeavors unintentionally highlighted the profound bureaucratic and economic constraints faced even by those tasked with modeling how to build. For example, the University of Dar es Salaam (UDSM) Building Research Unit produced a film called *Jijenge Nyumba Bora* (Build a better house) in 1977 to demonstrate the skills and process of home building from start to finish. Before they even began, they first struggled to find where to build. After realizing how long the waiting list for a planned plot in the city was, they built instead in Manzese, one of the city's largest "unplanned" neighborhoods. Their next obstacle was locating enough cement. Cement prices were officially controlled by the state, with prices published in the newspapers and aired on the radio. The team set out to buy the ninety bags they needed to build the house at the official price of 18 shillings per bag. They quickly learned (or likely already knew) that this was impossible. Due to shortages and subsequent hoarding and racketeering operations, no one in Dar could buy significant amounts of cement for its official price. UDSM's team could only buy twelve at close to the set price. The rest they obtained only after a lengthy process of black market negotiation. Even at stores where there were seemingly good stocks of cement, store owners warily eyed customers and told them the official price. However, "if you give them money for let us say 10 bags . . . you will receive an answer that the whole lot of cement is already sold and is just here waiting for the buyer to collect. There is no cement for sale at present." Only after these initial interactions when the store owner felt the customer was "not an alarming person" would they offer to sell the cement but at a much higher price: "If you agree, you pay and collect the cement immediately. Sometimes the store is closed and you assume that there is no cement. Very often however you can see a person approaching, telling you that there is cement, but the price is 27 and no receipt will be issued."[106]

There was clearly a delicate (and illegal) terrain that builders navigated in order to procure cement, and it was not over once it was purchased. Because cement was so scarce and often officially rationed, when someone did get their hands on a few bags it easily attracted unwanted attention. For example, a university professor I interviewed who was living on the campus property at the time began collecting cement with plans to build a home. Before accumulating enough, he ended up moving from campus because buying and bringing in bags of cement had led others to be suspicious of whether he was hoarding it, which was a crime. He then moved to the nearby new neighborhood of Makongo Juu, where he had acquired a piece of land from a local elder. There, away from colleagues' eyes and assessment, he began to slowly accumulate enough to assemble his house. Mariam Hassan, building on the other side of town in Mbagala, recalled how she had to be very careful storing cement bags as she accumulated enough to build a house because they would otherwise be stolen. Hamada Ali Mnora, who worked at the Rafiki Textile Mill, also took another job to save money for cement. His building process was punctuated by the fits and starts of saving enough money to buy cement for making his own cement blocks.[107] Another resident in Mbagala recalled exchanging food for building materials when he did not have enough money to buy them.[108]

The slow accumulation of cement was not a new method of building brought on by resource scarcities of the 1970s, though the circumstances exacerbated it. Rather, this slow accumulation of resources mirrored the manner in which urban communities had been building for decades, and highlights what had always rung particularly false in the colonial and postcolonial state's Manichean categories of building materials. Indeed, this incremental process of home construction still shapes building across the continent and the developing world. Building is accretionary and best seen as a collaboration between temporary and permanent materials, rural and urban economies, and a process of negotiating between future aspirations and current resources. Houses are frequently transitional spaces, and building homes was not about adherence to materials but tactical and timed bets on the future. As Leslie wrote of Magomeni in the late 1950s, construction increased "whenever money becomes available; a good rice crop on the Rufiji [a river delta in the nearby coastal region] will be reflected in a crop of iron roofs in Magomeni; a cashew harvest in Kilwa brings a number of houses up into the cement-floored class."[109] This kind of building in Kiswahili is called *humo kwa humo*, or side by side. A 1977 World Bank Sites and Services study demonstrates the resulting bricolage that made up not just individual houses but the larger landscape of the city: in Dar, 50 percent of houses had concrete floors while the other half had earth; 67 percent had walls of mud and poles, 30 percent had sand cement.[110] For most in the 1970s and early 1980s, there existed in reality no either/or categories of permanent and impermanent materials: concrete, mud, sand, thatch, and brick existed in concert with each other—side by side—replacing, overlapping, and repairing over the course of a house's life. Houses were themselves processes and liminal assemblages as was the larger cityscape.

Just as materials were a mixture of available resources, Swahili-style houses also cultivated a variety of ad hoc assemblages inside their walls as well. The Swahili style has likely remained the predominant mode of building for as long as it has not merely because it is an enduring local tradition; as noted earlier, these are incredibly flexible four- to six-room structures, holding both strangers and families together, renters and landlords alike. The state had realized this in the 1960s, when the first new NHC houses were built during the era of slum clearances. Swahili-style houses made with concrete blocks and corrugated iron roofs were "conceptually and practically" the main form of housing built by the state until 1971.[111] Those who bought these homes (others remained the property of the state) could live only in one room and were required by the NHC to rent out the rest, with part of the rental income going to paying back the loan.

By the 1970s, though, when the state took aim at landlords, they also started building single-family homes instead, severely crippling their ability to provide sufficient housing for the growing city. Urbanites building for themselves, however, did not abandon the Swahili style. With as many as one family per room, landlords could get 23 percent gross annual return on their investment.[112] It was a hugely profitable enterprise and the laws against landlords were far more focused on a rentier class in downtown Dar than on the modest houses expanding into its periphery. Owners could use rent to acquire more "permanent" materials over time, guaranteeing them ultimately more return on their capital. Renters could serve as guardians, protecting the property and investments against theft as it underwent its material transformations. Finding a room to rent for your family was generally cheap. Cousins and brothers could come from rural areas and take advantage of the room rented by another family member until they got on their feet. Men whose wives stayed in the hinterlands during the week and farmed could find a room closer to where they worked. Couples might rent as they endured the sometimes multiple-year process of securing land and materials of their own. Despite the relative cheapness of land during the 1970s, the scarcity of building materials made building an incremental and expensive process; only 17 percent of people who lived in unplanned areas owned their homes.[113]

To conclude, the city's built environment never reflected dogmatic preferences or categories but was far more legible in different ways. Houses and neighborhoods could reveal personal and national histories or fortunes: a family's current employment, how long they had been in the city, whether they feared removal by the state, the result of a good harvest, and the availability and cost of permanent materials. And yet humo kwa humo was also not divorced from the state's material categories; they remained in constant interplay. Humo kwa humo was reactionary in that it evolved out of circumstance but also because it could be a clever legal and political response to insecurity. Indeed, the colonial government, in prohibiting building in permanent materials outside of planned neighborhoods, knew this was the process of building. They knew houses made of permanent materials would be harder to remove and more expensive to compensate when the extension of the city's "planning zone" called for their destruction. The end of slum clearance schemes in 1972 also was an acknowledgment that "improvement" was the method of building by most urban Tanzanians and that other methods were not realistically possible without cheaper permanent materials. But if the goal was to repatriate jobless urbanites to the country, a permissive stance on permanent building could also make eviction harder.[114]

The slow construction of houses in the 1970s also suggests that the state's push for bricks had failed in urban areas. Nyerere implored citizens to get over their "mental block" against bricks in 1977 because so many were living in unfinished houses, or renting elsewhere while slowly building as they could get cement. Attempting to again promote bricks in 1982 in the *Daily News*, Kilimwiko wrote that waiting for cement, corrugated iron sheets, and nails led to the "omnipresence of somber looking unfinished buildings in most cities and towns."[115] But for those with little security in the city that was already so insecure, the process of building itself could buy time, particularly because use and improvement were frequently the moral and legal determinants of rights to land during the socialist period. Anthropologist Kelly Askew has observed that in postsocialist Tanzania, where the same land tenure laws prevailed, the cityscape was still marked by "building at a glacial speed" as a strategy of the poor. "Seeing so many incomplete houses," Askew writes, "I had always attributed it to lack of resources." But while cost certainly factored in, Askew also realized that adding minimally to a house "under construction" every year allowed families to prove to government surveyors that "work was ongoing and that the property was indeed being used."[116]

This strategy, common in many parts of the Global South, is a way to secure space in the city until a time when building is more feasible. Cement is particularly well suited for this kind of building. In places where mortgages cannot be obtained, construction lurches and lunges with periods of disposable income; families can accumulate cement slowly until they have enough time and materials to build. As Forty observes of this common tactic, "incompletion becomes a permanent state."[117]

Highlighting these ongoing states of transition rather than the discrete transformations that postcolonial states frequently imagined as the end result of decolonization brings to mind Tim Ingold's work on dwelling. Ingold writes that "environments are never complete but are continually under construction."[118] Ingold identifies two perspectives on home making: a "building" perspective and a "dwelling perspective." In the building perspective, "worlds are made before they are lived in."[119] This perspective embodies the perspective of the architect or planner: "first plan and build the houses, then import the people to occupy them."[120] And indeed this is also generally a statist perspective. The dwelling perspective acknowledges that building "is a process that is continually going on, for as long as people dwell in an environment. It does not begin here, with a pre-formed plan, and end there, with a finished artifact."[121] African cities are easy examples to point out as landscapes in process, but I suggest that they offer to urban environmental history a particular imperative to highlight process and to continue to think of unfolding environments and landscapes rather than points of departure and arrival. This disrupts teleological state narratives of nation building because indeed they *are* always disrupted as they unfold. It is also worth noting that in key ways that frequently went unnoticed by the state, humo kwa humo perhaps best articulated many of the modernizing aspirations of both colonial and postcolonial research into building materials: this slow accretion of resources was utterly future-oriented, modular, and repairable, and used what was locally available, whether that be mud and thatch, coral, concrete, bricks, or aluminum siding.

Chapter 4

WAITING

Loitering, Laboring, and Breakdown in the City

"Amka, amka, amka!"

At the end of August 1978 someone under the pseudonym "Disturbed Sleeper" wrote into the *Daily News* to complain about the call, "Amka, amka, amka!," heard "almost every day in the morning."[1] The letter writer was referring to the call early each workday over the airwaves of Radio Tanzania Dar es Salaam for workers to "wake up, wake up, wake up!" These calls were accompanied by a cacophony of banging and knocking intended to jolt the late worker from his or her slumber. "Disturbed Sleeper" reasoned that this was not necessary: "At this stage of our development, it is impossible that one should not realize the importance of work as well as one's responsibility to the nation." An epistolary conversation unfolded in the next few weeks in the newspaper's Letters to the Editor section discussing whether "*Ndugu* [comrade] Disturbed Sleeper" was correct. Mja Maa from the University of Dar es Salaam wrote back arguing that *Amka* was "revolutionary" and that at no stage of development can "one can boast that he is well developed and need not to be explained about work."[2] Others chimed in that if there was going to be a call to work, it should be accompanied not by loud banging but by an "educative" message: "We don't have to oppress ourselves psychologically in the name of revolution."[3]

The radio's playful but obnoxious clanging perhaps sounded a more pervasive anxiety regarding the work ethic of urban Tanzanians. The daily call to get out of bed was just one of many persistent reminders that the state saw urbanites as lacking urgency. Lurking in the newspaper exchange over "Amka, amka, amka!" was the persistent fear that a late worker would lead to a forestalled Tanzanian future. Even in the most domestic of settings, Tanzanians were constantly being reminded, in Nyerere's famous phrase, that "we must run while they walk."[4] And yet, according to "Disturbed Sleeper," the city's workers did not need to be reminded of the importance of hard work. As this chapter will explore, those who were utterly anxious to work hard did not want an alarm but rather a readily functioning public transportation system. Arriving late to work, they argued, was not from sleeping too long but instead due to the city's struggling public buses: it was the state of urban infrastructure, not personal character, that imperiled postcolonial nation building. Taking a cue from the radio alarm, this chapter is not actually about punctuality and hard work so much as the anxiety of idleness lurking behind the daily clanging over the airwaves: the fears that Tanzanian progress would be stymied by waiting, breakdown, and the suspended animation of city life.

Urbanites in the 1970s spent a lot of time waiting. But depending on who you were or who you asked, that act of waiting was called different things and implicated various culprits and prescriptive fixes. For example, the state's word for unmoving bodies in the city's landscape was loitering.[5] Loitering was a moral problem that emerged mostly in urban environments. Indeed, loitering was even understood to be *caused* by cities inasmuch as cities were a haven for such behavior. The legal category of loitering made idling in public spaces criminal as well as immoral and indeed the government even waged a war against "idlers" in 1976 to clean up the economy and city. For President Nyerere, idlers were holding back the potential of ujamaa.

Having long lived with the awareness of state attitudes toward loitering, urbanites in turn sought to rhetorically reframe their own frustrating periods of waiting in the public spaces of the city as the result of failed urban infrastructure. Their grievances took the form of letters to the editor, but also bodies in space: what counted as work, loitering, or waiting was negotiated and contested in the public spaces of Dar es Salaam, in queues at the bus stop, and walking rather than riding along the city's roads. These opposing discourses of Dar as home to irresponsible idlers (*wahuni*) or impatient workers (*wafanyakazi*) animated a public conversation in the 1970s on urban ujamaa, productivity, class, and the changing nature of labor. Whereas the previous chapter considered some of the environmental and material contingencies of nation building, here my focus is on how different temporalities of waiting for the bus, waiting for work, waiting for spare parts, and waiting for a particular future collided on Dar's streets and in the pages of its newspapers.

But beyond apparently shaping the moral constitution of Tanzanians, waiting also shaped urban environments and infrastructures. In the final section of this chapter I examine how roads, buses, and their awaiting passengers created landscapes of deferral. These acts of waiting highlight the discordant temporalities of living in the city and its daily rhythms of care and repair. And for all its presumptive inaction, moments of waiting emerge as a catalytic force shaping the urban landscape in the socialist era. Drawing attention to waiting also picks up on what Ato Quayson argues is the "ultimate challenge in describing the African city." By what methodology, Quayson asks, can we "integrate the study of assumed ephemera" of the urban experience "into our interpretations of material conditions and how we read the two domains for insights into processes and structure"?[6] Examining urban environments as emerging from moments of waiting and suspended animation considers the lasting effects of these ephemeral moments. In the stalled-out moments of "breakdown" when urban infrastructures grind to a halt, city landscapes took shape anyway, as forces of nature and frustrated urban communities refused to wait. Taken together, ephemeral seasons of sweaty afternoons at bus stops were not just fleeting moments of discomfort but experiences that gave shape Dar's environment and cityscape.

Loitering in Colonial Dar es Salaam

Paradoxically, while defined by not moving, loitering emerges in colonial Tanganyika out of a fear of African mobility.[7] Loitering can only exist after the rules of how public urban space should be used and by whom have been established. To repurpose Mary Douglas's famous definition of dirt, loiterers are people out of place.[8] Like dirt, standing around or waiting for something was not intrinsically problematic, but certain landscapes and times of day made it so. And while it is technically a condition anyone might find themselves in, those accused of loitering in colonial Dar were nearly always African men who came to the city without work or remained after they no longer had it.[9] Inscribed in law, the colonial state used the category of loitering as a tool of social control not only to preserve the racial division of city neighborhoods but to separate the growing class of African workers in the city from those seen by the state as vagrants. Policing supposed loiterers accomplished three things: it captured labor by deporting jobless men to rural areas for work, it protected laborers from contact with "loitering" classes, and it preserved a particular (European) ideal of how to occupy public urban space.

Following World War I and the collapse of German authority in Dar es Salaam, the incoming British administration feared that the city had become a lawless place. To control movement in and out of the city, the anxious authorities implemented a "Destitute Persons Ordinance" that allowed the colonial state to remove anyone one from the city found without employment or unable to demonstrate their means of subsistence.[10] The British worried about how they would house young men arriving for work in order to keep them separate from "groups of native loafers sprawling across our pavements in main streets as well as in others, coughing and spitting . . . and occasionally gambling."[11] When the labor commissioner in 1928 discussed establishing a labor camp for some of the city's workers, they rationalized that with proper fencing and the right location, they could keep young men separate from the "native town" so as "to diminish the likelihood of town loafers infesting it." The administration wanted to enforce this separation because they feared that "workers" would become "loiterers" if exposed to such elements, and yet also saw them as mutually exclusive categories.[12] By the 1930s the police commissioner had inaugurated the first forced removals of "unemployed" Africans from Dar to the countryside. Headmen were to be appointed to each street "to know what is going on his street, particularly to note the arrival of strangers."[13]

Many of the supposed loafers who called Dar their home were not in reality unemployed but rather engaged in work that simply did not benefit the colonial state. To rein in these forms of labor urban administrators implemented sanitation laws and urban planning ordinances to curb activities such as brewing alcohol, hawking fish, or selling tea.[14] By the end of the colonial period, purges of loiterers had become a "daily occurrence."[15] By 1958 Africans who wished to remain in the city were required to have obtained a travel permit at the district office certifying their employment and proof of residence in the city for the past four out of five years.[16] The state had also transformed overt control into the urban entitlement of middle-class suburbs: the best way to ward off deportation was a concrete house, a wage, and a nuclear family.

A Restless New Nation

With independence, the state's hostility toward loitering did not diminish. Many urban elites also supported policing public space as a strategy for controlling and cleaning up an increasingly crowded city. But much like slum clearance schemes, the postcolonial state had to carefully articulate its position on loitering in a moral register that aligned with its mission to improve the lives of the country's poorest. For President Nyerere, loitering was the result of capitalism and colonialism: two systems that purposely produced surplus idle populations. Loiterers were thus troublesome because men and women who were not working were inherently exploiting someone else who was compensating for their laziness. A son sitting idly in a city was relying on his country family to farm both for his consumption and to support the national economy. Nyerere also carefully framed any efforts to remove the urban poor with his concurrent efforts to nationalize the property and businesses of the rich, which he saw as forms of accumulation that also promoted idleness among the privileged.[17] Thomas Molony traces Nyerere's intellectual ire for loitering to his time in Ralph Piddington's social anthropology course in Edinburgh. Reading Piddington's book, *Introduction to Social Anthropology*, a young Nyerere underlined an example of the Wahehe in Tanzania and their use of cooperative labor. These anecdotes, Molony argues, helped develop Nyerere's conception of "African socialism" as cooperative and traditional. They also shaped his notion of what he came to call the parasite: "The laggard—or 'the loiterer, or idler'—later became one of Nyerere's favourite targets, a 'modern parasite' who he claimed did not exist in his traditional Africa, where 'loitering was an unthinkable disgrace.'"[18]

While Nyerere's conception of loitering was clearly tied to his understanding of the transformative potential of African socialism, it nevertheless echoed a colonial rhetoric of policing the uses of urban space: "Nobody should go and stay for a long time with his relative, doing no work because in doing so he will be exploiting his relative. Likewise, nobody should be allowed to loiter in towns or villages without doing work which would enable him to be self-reliant without exploiting his relatives."[19] Cities and towns were once again drawn up as predatory to rural production but now a loiterer was not just taking advantage of a family member but failing the new nation.

When read alongside postwar ideologies of development and modernization, eradicating loiterers was also tied to globally circulating ideologies of speed and an acute awareness of time passing after decolonization. As several scholars have recently elucidated, Walter Rostow's theory of "stages of economic growth" became a central text as well as metaphor for development for the newly decolonized world.[20] Rostow's theory entangled notions of national development and underdevelopment with emerging ideas in psychology that marked human progression in similar concrete stages. New nations were the equivalent of newborn babies: they faced the challenge of moving as quickly as possible through inescapable and linear phases of development that the industrialized world had already passed through.[21] And if development was linear, then speed was the sign of national success.[22] Keenly aware of this, leaders across the Third World were anxious to get out of the "waiting room of history."[23] Nyerere, for example, noted that "while others are going to the moon we are trying to reach the village and it is getting farther and farther away."[24] By framing Tanzanian development goals as modest (the village, not the moon) Nyerere made speed and swiftness all the more urgent.

By the 1980s, however, Nyerere was reconsidering what "catching up" should look like for Tanzania, and reappraising the Third World's desire to emulate Western progress. By falling victim to "buying the most elaborate building and the latest invention in every field regardless of our capacity to pay for it [or] even to maintain it," Nyerere worried that Tanzania had created a "continuing dependency on the importation of technology which then requires us to produce for export, regardless of our peoples' present hunger and present needs."[25] I will discuss the material dimensions of this effort to rethink technological intervention in the next chapter, but it also had significant labor implications for workers, just like brickmaking. By turning away from imported technological fixes, the willingness of Tanzanians to undertake labor-intensive solutions that did not require much capital became the nation's primary resource. Nyerere's skepticism was not merely ideological but also based on Tanzania's ongoing experience in staving off breakdown in industrializing sites like the Wazo Hill cement plant. "Catching up" had become rife with instances of stalling out, and long periods of waiting for supplies, expertise, and maintenance. Likewise, other socialist nations who similarly sought to distance themselves from a narrative of capitalist progress also began articulating progress in different (yet still linear) terms.[26] The president urged comparisons not with the "latest inventions" of capitalism but with the ethic of hard work in other socialist countries. "TANU's language is now a language of revolution like that of China, Korea and Vietnam," declared Nyerere, "complete with revolutionary phrases. The Chinese are surprised when they come to our country. The land is fertile but her people are only fond of revolutionary phrases. Think of the Vietnamese too. Where are they at this moment? All their able-bodied men and women are seriously at work."[27] This comparison to communist states reaffirmed the immorality of loitering as a particular socialist cause: hard work was the only way to escape waiting for technological salvation.

While Nyerere's rhetoric and campaign against loitering must be understood differently from that of the colonial administration, his interventions against *wahuni* nevertheless exploited the legal and social structure left over from the colonial period. As described in chapter 2, in the 1960s and increasingly during the 1970s, Nyerere began to round up "unemployed" urbanites and repatriate them into agricultural schemes much like the colonial state had done.[28] These campaigns effectively criminalized being in public places during the day. But, as many citizens anxiously pointed out, streets and street life were not just a tableau of the supposedly "unemployed." They also reflected a worsening economic crisis, the dwindling purchasing power of formal salaries that drove people to "informal" activities, and the failure of Dar es Salaam's public bus system.

By the final campaign against loiterers in 1983, known as Nguvu Kazi (hard work), the problematic class politics of capturing people on the street became impossible to overlook. The aim of the campaign, like many before, was to repatriate seemingly unemployed urbanites back into rural areas to work. Over five thousand "loiterers" were arrested in one of the first days of the campaign in the hours between ten o'clock and noon simply for being in the streets.[29] The ham-fisted enforcement of antiloitering laws had always clumsily hidden larger debates of who could be in the city and what defined urban work. Nyerere asserted that "no-one in this country has the right to wage-employment. What we do have a right to, is an opportunity to work. That opportunity exists. It exists on our land."[30] This definition of work was not tied to securing a wage and reflected a far more expansive notion of nation building, emphasizing rural development. And yet urbanites readily pointed out that, based on how antiloitering laws were enforced, officials implicitly left those with formal employment unmolested, assuming they were hard at work.[31] Thus these campaigns implicitly valorized wage labor.

As citizens took to the newspapers to complain about repatriation campaigns and failing urban infrastructures, they did not challenge the notion that loiterers were the enemies of progress. Rather, they contested what could be mistaken for loitering and, equally, what could be mistaken for work. Perhaps most loudly they rushed to point out how the city bus service routinely foiled their attempts to work hard rather than languish on the side of the road while those with private cars were left invulnerable to interrogation. The persecution of loiterers on the streets of the city revealed class divisions that were not supposed to exist in socialist Tanzania. As one citizen wrote to the newspaper: "I would like to question those in authority of their indiscriminate arrests of pedestrians and cyclists in the streets between 10:00AM and 2PM is legal and constitutional. . . . People using cars and trucks during the so-called work hours are not harassed. Why? Can't someone loiter with his car for petty personal business during work hours? Are we creating two classes of citizens in this country—the oppressed and the privileged?"[32] This free pass for the potentially lazy and unproductive manager or bureaucrat who might have privileged access to both a job and a car hit squarely in the center of the ongoing debate in Dar regarding transportation and the nature of socialism. Underlying the question of who could loiter, quite literally, was the poor state of Dar's roads and the ways in which roads shaped both settlement and mobility in the city.

Roads to Self-Reliance

Dar's basic layout comprises four major roads radiating out from the city center, with each arterial road reaching into the hinterlands and nearby rural regions.[33] With the birth of the British Dar es Salaam Motor Transport Company (DMT) in the late 1940s and the beginning of public transportation, it became easier to live farther from the city center. But formal road-building efforts persistently lagged behind the expansion of city neighborhoods. In the 1950s, the first major macadamized road connected Dar es Salaam to Morogoro, 120 miles to the west, with the other major spokes across the city soon following. Expanding informal neighborhoods hugged the sides of these crucial conduits as the city grew and spread. About ten kilometers from downtown along Morogoro road, the neighborhood of Ubungo grew from five thousand residents in 1967 to sixty thousand in twenty years.[34] After independence, these arterial roads remained some of the few paved roads in the city outside of downtown. Connecting the city and the port to the rest of Tanzania, nationalists imagined these roads as instruments for leveling the inequality of country and city created by colonialism's urban bias. The prominence of these major conduits has also meant that it was often easier to travel in and out of the city than to traverse its neighborhoods, which could be a tedious journey requiring the traveler to pass in and out of the center of the city.

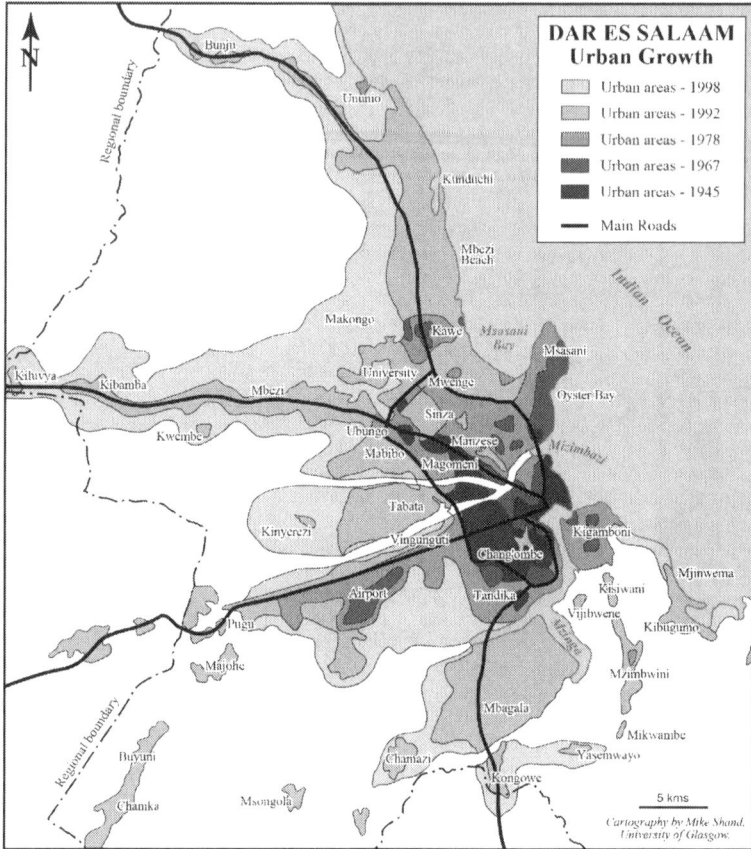

DAR ES SALAAM
Urban Growth

Urban areas - 1998
Urban areas - 1992
Urban areas - 1978
Urban areas - 1967
Urban areas - 1945

— Main Roads

After Briggs and Mwamfupe (2000)

4.1. Map of urban growth in Dar es Salaam 1945–1998. Map courtesy of Mike Shand, based on John Briggs and Davis Mwwwamfupe (2000).

Noting this emerging congestion, the city's 1968 master plan prioritized "open space, land use and traffic movement" to convey "the dignity and stature of the capital of a country which is playing a leading role in the development of Africa."[35] The plan envisioned a six-lane freeway that would promote growth in the north and south poles of the city. But after the national capital was relocated to Dodoma, the plans were shelved and the road never expanded. While urban Tanzanians have always conducted livelihoods between the city and country, by the 1980s the necessity of supplemental informal economic activities as well as alternative channels for food made mobility perhaps the single most important tactic for making their lives possible in the city. The sprawling boundaries and new uses of Dar's space reaffirmed the importance of automobility even as measures to conserve petroleum led to fewer cars on the roads and limited the days of the week private citizens could drive.[36] People needed buses to move across the city in the new ways their lives now required. This "uneconomic spreading of the city" placed great stress on public transport.[37]

Meanwhile, smaller roads that knitted together neighborhoods were not shaped by official road-building measures so much as by their everyday use. The effects of deferred maintenance worsened after the dissolution of the city council in 1973. Whereas the former city council had earmarked 7 million Tanzanian shillings (Tsh) for roads, the three districts that made up Dar after decentralization were earmarked only 700,000 Tsh for both maintenance and new projects.[38] Among other massive expenses, road building required a host of machines such as graders, tar boilers, tipper vehicles, concrete mixers, and the expensive importation of petroleum to make bitumen.[39] Following decentralization, most neighborhood roads were paved only if communities could generate their own funds. But without the construction of proper drainage systems, as one *Daily News* columnist pointed out, "we would be wasting public funds" to tarmac roads.[40] In essence, road building required that several other urban infrastructures be already in place, such as storm drains and sewerage. The roads of most neighborhoods, like Ubungo, Ilala, Kijitonyama, and Mburahati, were instead simply "paved by the tyres of the tenants who are to go and inhabit them; that is if they have car."[41]

Paving roads also required the reclamation of space; this meant that urban authorities would have to tear down homes, which took both monetary compensation and political will.[42] Also, when roads were paved, they quickly attracted new people to the area, leading to a new wave of transportation problems and the potential for overuse.[43] When major road projects were undertaken they frequently stalled out or inched along due to the vagaries of foreign exchange and shortages in spare parts. These delayed timelines caused construction prices to skyrocket, as they did when a thirteen-kilometer stretch of Pugu Road from the city center to the airport was first estimated to cost twenty-three million shillings in 1973 but just three years later and still unfinished, the new road had become a hundred-million-shilling project.[44]

The paucity of paved roads did not just determine the nature of morning commutes; it also shaped how neighborhoods accessed municipal provisions and infrastructures. Only along paved roads or in more established central neighborhoods could residents officially access electricity or get truck water delivery, let alone hook up to the pipes and infrastructures of the city's water or sewerage system. Sanitation trucks were also unable to travel along many of these smaller roads. Narrow streets and pathways necessitated specialist sets of jobs for the delivery of cooking charcoal and water and trash collection, by way of pushcarts. Thus, the very shortage of paved roads that may have kept people from work on time also offered up new opportunities for

4.2. DMT drivers in 1959 before nationalization of the company. Photo courtesy of Tanzania Information Services.

money making as residents worked to improvise and replace the absent urban infrastructures of their neighborhoods.[45]

By 1978 nearly half a million people lived beyond the tarmac in Dar.[46] The humid city was full of exhausted walkers, filtering out to the city's main conduits, where they would look up the road and hope for the bus. Dealing with the city's poor roads and public transport meant that waiting for the bus became a crucial shared point of view from which to consider both the city and urban ujamaa. Like the entitled car driver who escaped accusations of loitering, the waiting passenger became a ubiquitous character in the city's politics as well as landscape. Ever present, the waiting passenger could also far more easily be mistaken for a loiterer.

Liturgies of Waiting

In the early 1970s, around the same time that the cement plant at Wazo Hill was nationalized, the British Dar es Salaam Motor Transport became the parastatal Usafiri Dar es Salaam (UDA). While the state imagined that UDA would be key component of realizing urban socialism, it was unfortunately beset from the beginning by minimal investments and "low staff morale."[47] Tasked with providing cheap transportation, ticket prices never covered the cost of running the company. Particularly right after the oil crisis when a drought hit Tanzania and the state was forced to exhaust its foreign exchange funds on food imports, the UDA received only 35 percent of its requested budget. This left the company in a constant struggle to maintain its fleet of buses with necessary spare parts on hand.[48] As a result, even in its first few years of operation the UDA's services declined precipitously. By 1975, only 257 of UDA's fleet of 374 buses were serviceable and running. A decade later, the fleet had shrunk amid major urban growth, to just 205 buses with only 131 operational.[49] The UDA was persistently made to "cope with the rapidly changing conditions of a fast-growing city" as well as a struggling national economy.[50]

Despite these struggles, public transportation remained a crucial infrastructure for urban workers as was the case in many socialist countries.[51] Joshua Grace has argued that Dar's buses served as a microcosm for ujamaa itself: they were "transformative spaces that rallied strangers into daily conversations about their collective experiences and frustrations as residents of a socialist city."[52] Beyond these conversations that passed between passengers, the bus-riding public also articulated their frustrations in the pages of both English and Swahili newspapers spanning the decade after DMT's first troubles in the early 1970s. The act of sharing one's experience in the newspaper or in conversation while reading the plight of others also became a daily shared ritual of urban life among the newspaper-reading public, taking on an almost liturgical quality.[53] In this secular ritual of city making, these letters and articles describe the sensorial and corporeal experience of waiting and riding the bus and attempted to put this experience at the center of an ongoing discussion of urban work and nation building.[54]

These articles and letters published in newspapers ran the gamut of complaints: buses never came on time or if they did they were too full for passengers to even stand. Buses would pass stops without slowing down and being a passenger was dangerous, while being a pedestrian or a bicyclist frequently was more so.[55] People wrote to complain about rude or lazy drivers who betrayed Tanzania's socialist ethos by keeping their buses running while they took breaks, and wasted precious petrol.[56] Surely hoping the DMT management were also reading the newspapers, some letters urged readers to visit neighborhoods and observe how long the *foleni* (lines) of workers waiting for the bus were.[57] Writers/riders also complained about buses falling apart, the lack of spare parts to fix them, and the high cost of poorly maintained potholed roads sending newly repaired buses back to the shop to wait again for the arrival of more spare parts.[58]

While the letters covered a range of topics, they frequently took on a similar form of asking readers to imagine the hardship of trying to get to work, calling on empathy and constructing a collective understanding of the corporeal and psychological discomfort of waiting. When read together, these letters are one of the most evocative sources for understanding how urbanites experienced and shaped Dar's urban landscape in this period. A bird's-eye

4.3. A long line of passengers waiting for the bus, with the headline "DMT What's Up?" Published in both the *Daily News* and the monthly news magazine *Nchi Yetu* in October 1973.

map of the city's roads cannot convey how the stuttering service of the bus and the cascading consequences of failing to get on the bus shaped Dar's streets, neighborhoods, and livelihoods. Waiting was both a personal griev- ance and a shared plight. As if in reply to Nyerere's command that Tanzanians must "run" while others walked, one newspaper columnist opined, "We used to move quickly, but this is not the case today." This same article represents one of the more epic examples of the genre, charting the daily travails of a commuter named Ramadhani Sefu. From his home in Manzese to his job as a government clerk at the waterfront, Sefu's journey is retold in excruciating detail, conveying the exhaustion of his daily trip:

> He wakes up at 5 am and by 5.30 am he is at the nearest bus stop waiting for transport to take him to the city centre. Several people are already there. A bus approaches from town heading towards Ubungo. Ramadhani and his colleagues watch it as it passes them. They now wait for it after picking people at Ubungo. The sun rises from the East and a bus approaches from Ubungo towards the town. It is fully loaded and does not stop at Manzese. The second and third bus pass by too. Some waiting passengers decide to walk—some towards Ubungo to catch the bus there and others towards Magomeni, hoping to get one there or walk to their destinations. Sefu remains at his Manzese bus-stop. It is 6:30 am and there are many people there. And the buses just pass by. At last, one stops but it is almost full. A few people manage to squeeze in, mostly teenagers. Sefu cannot even get near the door.[59]

Sefu's struggle to get to work is largely one of not getting on the bus: a competition with other would-be passengers, particularly with those more physically aggressive. He must then explain his lateness to work to his boss, who has a car and thus would not understand. The reverberating effects of waiting echo through his day and limit his ability to be a productive worker. On his way home, he faces the same dilemma, arriving home "tired and hungry."

In another article calling for the DMT to overhaul their buses, the appeal is structured similarly to Sefu's tale. This man left his house early "but he will not get a bus until an hour later. Such waiting eats on the energy of the worker." Personal fatigue in this way becomes political. The writer suggests that "we have got to go further and investigate what workers and employers lose as a result of this time wastage." Someone who faces waiting for the bus only to travel on it standing up and packed in "cannot be expected to be very productive. His stamina will have been very much depleted. His mental composure and peace will have been upset and he would be demoralized. Result: a short and wasted working day."[60] The high price of public transportation paid by Dar's workers was not just the fare but also the investment of time and prospect of losing one's job. These public complaints sought to explain lateness and lack of productivity as the result of infrastructural breakdown rather than a personal or moral failing.

While these liturgies of waiting captured a universal frustration of riding the bus, there were still clear dividing lines across these experiences. Despite the emergence of strict rules for how to queue for buses, many stops still dissolved into physical struggles to find room when one arrived. When a full bus showed up at a stop, would-be passengers rushed to climb in (and out) of windows just to find a seat. Such scrambles favored certain riders over others. Women I interviewed, for example, recalled being scared once the bus arrived as people fought for seats. To avoid the scuffle, they waited longer or took off on foot.[61] Beyond the physical vulnerability of finding a seat, women also faced potential violence by simply occupying outdoor public spaces. This stretched back into the 1960s and earlier campaigns to discipline urbanites and city spaces. As Andrew Ivaska writes, the Tanzanian Youth League inaugurated a campaign in 1968 called Operation Vijana, which was aimed mainly at reforming women's dress to evoke Tanzanian rather than "Western" values. The most intense and violent confrontations between enforcing young men and targeted women happened on "bus routes, bus stations and downtown streets."[62] Cloaked as assaults on women's sartorial choice of miniskirts, the public humiliation of women were struggles over who could claim space in the city as well as access to wage labor.

Bus riding then was itself a privileged, if ephemeral, position. Imagined to be the right of workers, it was a form of city dwelling that favored tough young men. The act of writing to newspapers about public transit doubled as a conscious act of claiming citizenship in the city through a seat on the bus.

4.4. Painting by Mohamed Raza depicting the struggles to get on the bus. Published in the *Daily News*, March 26, 1983.

Since the colonial era, invoking a right to the city had been incumbent on se-curing labor; having a job imbued the letter writer with a sense of importance and urgency that could distinguish oneself from a loiterer. Whereas loiter-ing was framed as a volitional act of wasting time (and thus subject to moral interrogation), these liturgies of waiting implicated factors beyond the con-trol of the passenger. Letter writing separated their authors from those who supposedly languished in public spaces while shirking their nation-building responsibilities. At the very least these letters asked other members of the newspaper-reading public to bear witness to their foiled attempts to labor.

The Guidelines

As noted earlier, seething just beneath the daily frustration of waiting for a seat on the bus was the issue of who had access to private cars. Waiting workers never failed to notice managers driving home company cars and thus pointing out the unfulfilled promise of equality between the worker and bu-reaucrat under socialism. What brought these complaints t
o the surface was actually a government policy document published in the early 1970s regarding the rights of workers under ujamaa. *Mwongozo wa TANU* (The party guidelines of TANU), published in 1971, was written in the wake of both the overthrow of Obote's government in Uganda and the invasion of Guinea by Portuguese mercenaries at a time when the state was nervous about both foreign and internal sabotage. *Mwongozo* sought to em-phasize the important role of workers in securing the future of Tanzanian so-

cialism in tandem with their rural counterparts. These new guidelines urged factory workers in particular to stay the course of socialism in Tanzania and avoid temptation to stray to new leaders or ideals.[63]

Mwongozo's second goal and broad context was to reassert Tanzania's sovereignty by warning against both unwelcome imperial impositions from outside the new nation's borders and counterrevolution from within. These potential impositions, though, were not strictly political but in fact more broadly economic incursions that prompted a series of nationalizations of industry and property. To protect against the power of foreign corporations, the state urged workers and managers to be vigilant regarding "buying items that do not help our economy," and emphasized the necessity for importing agencies and parastatals to follow strict guidelines to help realize the nation's socialist goals. In encouraging this rededication to their labor, *Mwongozo* asked workers to actively question under what systems and logics workplaces operated. As a microcosm of the nation, workers should steel their workplaces against both foreign ideological penetration and the "oppressive habits" of bosses who still ruled with a colonial mentality.[64] Thus, the state's own labor policy explicitly connected a commitment to work with being vigilant against the abuses of the managerial classes.[65]

If *Mwongozo* announced that workers' rights were now equal to those of bosses and bureaucrats, this was both an empowering rhetoric and a confusing directive. How exactly were workers to keep managers in check? How would the state support them in these efforts? *Mwongozo*'s only real directions were that the path to this socialist future would be hard won through unlearning "colonial working habits": "We have inherited in the government, industries, and other institutions the habit in which one man gives orders and the rest simply obey them," note the guidelines. "The result is to make them feel a national institution is not theirs, and consequently workers adopt the habits of hired employees."[66]

Armed with the new language, employees wasted little time before airing long-festering workplace complaints. Michaela von Freyhold, in her 1970s study of Tanzanian industrial workers, describes how quickly workers invoked the political vocabulary of TANU to "articulate their feelings and opinions in ways which their superiors often find difficult to match." If a manager could not muster a reply, workers would claim "he has got no politics" (*haina siasa*), implying not a dearth of deeply held ideology so much as someone lacking the "art of manipulation."[67] Likewise, *Mwongozo*'s hasty embrace by workers did not necessarily reflect their embrace of party ideology. Rather, the document provided language and permission for workers to question their exploitation, whether under capitalism or socialism. Workers had recourse to other livelihoods through "economies of affection" and thus were powerfully positioned to protest working conditions.[68] More likely, workers hoped that "they might save enough money to set themselves up as farmers, independent artisans or business men" but "'Nobody is born to die in a factory,' they sometimes say."[69]

After *Mwongozo* was published, the early 1970s was marked by a series of workers' strikes. This included the "downing of tools," locking out managers, and occupying factories as workers demanded more respect, higher wages, and frequently complained that they faced racism from foreign managers.[70] As soon as workers started airing their complaints, issues of transportation became a central grievance. Workers in Dar seized on mobility as an aspect of workplace politics. In an unintentionally apt turn of phrase, Issa Shivji writes that *Mwongozo* "acted like a vehicle to carry the contradiction between the workers and the bureaucratic bourgeoisie to the fore."[71] Workers took note when managers abused their access to company cars, using them to run personal errands, or to go back and forth to plots of land outside of town and travel to rural areas. One worker wrote to the *Daily News*, "I am greatly concerned with those who are provided with transport amenities in that they are people who earn eight to fifteen times the MINIMUM WAGES. Sometimes they have their own cars—and yet we give them more to make them more productive!!"[72]

Tying their struggles in getting to work and their ability to labor productively to these unfair bureaucratic advantages, workers took action. In one example, Mr. Kashajia, the personnel manager of the American Tobacco Company of Tanzania, arrived at the factory in May of 1973 to find a workers' strike. He was ejected from the premises until the police arrived, armed with tear gas. When the event came before the Permanent Labour Tribunal, one of the major grievances workers cited against Mr. Kashajia was that he had driven the company's Range Rover to attend his father's burial in Bukoba, costing the company 5,820 shillings.[73] At the Bahari Beach Hotel workers did not actually strike but rather refused to be fired after they were deemed "redundant." Coming back to work daily, they argued that the company must have money to pay them minimum wage if they were spending five thousand shillings a day on "saloon cars" for top officials. At Kioo Limited, a glass factory, workers also demanded to know why, when *Mwongozo* "says we are equal" twenty managers had ten saloon cars at their disposal while workers had only one bus for morning transport and nothing in the evenings.[74]

These worker complaints suggest that not only were managers inhabiting a different world of privilege at work, they were also navigating a very different city outside the office and factory. Bureaucrats and managers could come and go while workers felt stuck in place. By withdrawing their labor and making factories—and the national economy—wait, workers placed their frustrations with immobility alongside national ideologies of progress and speed. These episodes suggest that workers focused on preserving their access to a broad set of opportunities that shaped the advantages of urban life and instrumentalized work for those purposes. They readily connected the material conditions of living (and waiting) in the city to the conditions of their labor. Their work environment was inherently about the city's environment and infrastructure too.

In some cases, worker protest over transportation was successful and companies purchased buses to compensate for the failure of the public transportation system.[75] State officials, however, feared that private buses would entrench class divisions in Tanzanian society. In August of 1974, TANU prohibited private bus purchases while also introducing a fare hike on public buses. Announcing the new ban, the minister for communications, Job Lusinde, dismissed the idea that transportation affected the productivity of workers, citing the Friendship Textile Mill as an example of a productive industry without private buses.[76] And yet earlier the same year the prime minister, Rashidi Kawawa, had asked the National Bank of Commerce to provide buses for their employees explicitly to increase productivity. The *Daily News* meanwhile published two editorials arguing that private buses would "reinforce disparity between town and countryside—an anachronism which the Party's policy of socialism and self-reliance repudiates."[77] For workers, though, a gap already existed. The gap was not between city and country but between worker and manager: "Those officers who banned private buses do not use public transport. They have personal or official cars and have no idea of the sufferings of UDA passengers."[78]

The state backed down from their prohibition just a month later, likely aware of how deeply unpopular and potentially troubling this ban was. By the 1980s the central government ministries and departments had 29 of their own staff buses and a total of 438 vehicles. If even just 70 percent of those were in good enough shape to drive on Dar's roads, that fleet reached nearly 90 percent of the capacity of Dar's public bus company.[79] In 1983, when city bus services were still under strain, the state enlisted these private buses to operate alongside their own buses to improve the city's transport. Fifty-seven parastatal buses were enrolled, charging the same fares as public buses.[80]

An illegal private sector of buses and cars also emerged on the streets of Dar es Salaam in the 1970s.[81] These were usually cheap, rebodied Korean trucks that had been turned into passenger vehicles, and they remained illegal until 1983. Some people with smaller cars and trucks also began functioning as informal taxis. A university professor recalled to me how buying a truck at one point allowed him to not only wake up early each morning and distribute meat to butcheries but also transport workers to their jobs afterward. "I was able to make more by seven in the morning than I did in my daily wage as a university professor."[82] By the 1980s these makeshift private buses bore the name now used for public buses today: *dala dala*s, a reference to the five-shilling coin it took to ride them, which was colloquially called a *dala*, a swahilization of the word *dollar*.[83]

While effective early on, striking died down by the mid-1970s, as it became clear in many confrontations that the state frequently did not take the worker's side. Freyhold suggests that workers realized that *Mwongozo* was not actually a call to radically alter the nature of urban wage labor. But without executing strikes and lockouts, workers still took part in "the perpetual go-slow which goes on in almost every industry."[84] Frustrated with working conditions but left with little recourse, workers settled on purposely stalling production, which became an enduring response to their own unaddressed frustrations with time and waiting.[85]

Waiting, Failure, and Recursion

As in many colonial and postcolonial cities, Dar's physical infrastructures "command[ed] a powerful presence." Their contingent and ephemeral functioning during the 1970s made them all the more visible and contestable. The breakdown of public transport in particular had clearly become a "means to critique the state and lament the failed promise of elites."[86] While the UDA certainly attempted to improve their services over the years, ongoing public complaints were not effective in cajoling the state as the owner of the parastatal into claiming responsibility for the diminishing productivity of workers. In part, this reflected how much the struggles of DMT were overdetermined: there were many potential sources of blame. The overdetermined nature of DMT's problems also made culpability a slippery affair. Waiting for the bus was part of a chain of reactions that could not just set off "cascading effects" but could spark a set of delays that would in turn worsen the original problem, sometimes to the point of near-constant failure.[87] In launching his war on loiterers in 1976, Nyerere took a moment to address those complaining about UDA's services. Like UDA's passengers, he also drew together loitering and public transportation—but not in a way that workers may have hoped. Nyerere deftly placed blame for UDA's problems at the feet of workers themselves. Poor bus service, Nyerere chastised them, was actually a symptom of Tanzanians *not* working hard enough rather than the impediment to hard work:

> All wealth comes from work. I do not want people complaining about UDA. How do you get UDA services without producing? UDA buses do not consume water—they consume fuel. Where do I get this fuel? I get it from a worker in an industry. He must work before I can get the fuel—not as a result of the UDA. I do not want complaints about UDA. UDA buses consume fuel, and we buy this fuel. We presently get it through the sales of cotton, cashew nuts and other crops. I should not be forced to import *khangas* [cloth] and cement because we produce them so that we can import fuel for the little money we have."[88]

As Nyerere frequently did, he did not deny the challenges or problems facing Tanzanians but rather urged every "worker" to understand their role in protecting the fragile and contingent nature of the nation's economy. For him, workers were foolishly missing the point that their diligent labor was what would allow the state to fix the transportation system. When workers failed to show up to work and blamed their absence on the UDA, it meant precious foreign exchange would have to be spent buying overseas goods to substitute for those not made when Dar's factories were short of manpower.

My aim in exploring the specter of recursive failure here is certainly not to rehearse a pessimistic narrative of postcolonial African economies and infrastructures. Rather, I am interested in how perilously *interdependent* infrastructural "breakdown" was for a new nation and how conscious of this reality both workers and the state were. A bus breaking down happens in every city. But it sets in motion a different set of problems in a city where there is no backup fleet.

Waiting for the bus—which would then be overloaded when it arrived—was both a result of and the cause of recursive problems. An overloaded bus required more regular repairs and its weight could also damage unpaved roads. Once roads were in poor health, they would in turn also harm buses. Waiting for spare parts to fix these buses then meant longer waiting at stops and more crowded buses, reanimating the harmful cycle. The recursive nature of infrastructure and its breakdown endowed a sense of momentum to transportation in Dar, but rarely one of getting to your destination on time. Instead, it was a momentum that stymied the intentions of mechanics, operators, and passengers.

To dig into this recursiveness, one culprit at first glance was the very composition of UDA's fleet. The company had acquired their buses both through various state contracts and as part of foreign aid packages, and the resulting fleet was a patchwork of Hungarian, West German, Japanese, British, and Korean vehicles. Each had their own systems of maintenance and their own spare parts, sometimes requiring teams of foreign experts to repair them.[89] For a company that already functioned on just a third of its desired budget, this presented a logistical nightmare for keeping expensive spare parts in stock. Even relatively small breakdowns could send buses for long periods of convalescence in UDA garages waiting for the correct part to arrive.

With a decreasing fleet every year, Nyerere appealed to private companies in 1976 to donate their spare parts to UDA in a campaign called Operation Fufua Mabasi (Operation Bus Revival).[90] By 1982 UDA was even worse off, having lost over seventeen million shillings in the previous two years and having to ground large portions of its fleet simply for lack of tires and tubes.[91] With every passing day that grounded vehicles stayed in UDA's garages, they were also vulnerable to being stripped further for their parts and components.

Tires became so valuable on the black market that the UDA frequently could not even afford to replace them, grounding the buses for even longer. As a result, a fleet of ragtag illegal buses began running "routes normally reserved for UDA, cashing in on the latter's inability to acquire tyres and tubes."[92] And while buses and cars were seen as antagonistic categories of transit, they were also interdependent. As cars aged and went unreplaced due to import bans on personal sedans after 1978, more people relied on city bus services.

To dig in even deeper, simply obtaining spare parts was also more complex than having enough foreign exchange to order them. Shipments of spare parts could arrive in Dar's port facilities and sit for weeks or months before being sent on to their destination. Dar's port was first expanded in 1956 to facilitate the Tanganyikan Groundnut Scheme, but in ten years' time it was already due for further expansion. Originally imagined as a major site of export, by the 1970s it functioned as a hub for the importation of an increasing list of goods, as well as the main port of the East African Harbor's Corporation. By 1975, port congestion worsened as oil imports alone had risen by 665 percent since its last expansion.[93]

Dwell time is the shipping term used to calculate and keep track of how long goods stay in a state of waiting between points of transport. Dar's excruciatingly long dwell time might have been troubling but it was not unique. For example, in the Lagos port in 1975 it could take three months from when a boat came into port for its goods to be unloaded at the dock. At one point over four hundred ships clogged the Nigerian port and perishable goods spoiled on board. "Much of the backlog consisted of cement," writes Michael Veal, "ordered by the government for various public works projects, and much of this was ultimately dumped into Lagos harbor, with the government accruing millions of Naira in demurrage fines on the remaining floating tons."[94] The singer Fela Kuti joked that instead of flying out of Lagos's notoriously chaotic airport, it was faster to simply jump from ship to ship all the way across the ocean.[95] Likewise in Dar, the functioning of the city's factories and parastatals were forced to accommodate long dwell times as goods were unloaded from ships.

Dar's port also served as landlocked Zambia's main conduit for imported goods and exported copper in the 1970s, allowing them to avoid relying on ports in white settler-ruled Southern Africa.[96] With this new responsibility, the efficient management of the port took on pan-African implications. A Zambian reporter showed up at the port in 1976 to see why tons of goods destined for his country had been marooned in Dar for months with no one there to even collect the goods. It became clear that some importers had concocted a scheme where they would "temporarily abandon their imports at the port only to buy them at a much cheaper price when the government was forced to auction them because of overstaying at the port."[97] Taking advantage of the increasing normalcy of waiting and long dwell times, buyers renegotiated their prices through exploiting the chaos caused by breakdown.[98]

It was becoming clear that waiting as an artifact of daily life generated both obstacles and opportunities. If breakdown became predictable, its consequences remained varied and evolving, leading ultimately to creative solutions.[99] Since infrastructures are attempts to rationalize both space and society, their failure prompts other "provisional" forms of ordering to emerge and take shape.[100] With no spare parts on hand, the necessity of re-pair sparked tinkering and innovation. The chronic ailing health of buses promoted the growth of informal labor servicing bus stops, buses, waiting riders, and in some cases directly feeding off UDA's failure by offering up other (illegal) alternatives. Brian Larkin writes on technology in Nigeria that parts sometimes have second, third, and fourth lives as they are reanimated through "the constant cycle of breakdown and repair." But incumbent in this resuscitation is "another waiting period, an often frustrating experience of duration brought about by the technology of speed itself." Because cars and buses were the prototypical technology of speed for most Tanzanians, they also embodied the accelerating future of Tanzania. Their proclivity to break-down magnified the frustrations of living between the "continually shifting states" of speed and stasis, expectations and reality.[101] Just as importers at the port found a way to capitalize on waiting, *bubu* garages—illegal and infor-mal mechanic shops—began to pop up at intersections under the shade of Dar's trees in response to the growing need to improvise spare parts and bus repairs.[102] Here, mechanics taught themselves and each other how to trans-form "motor vehicles from closed technologies produced in foreign factories to open technologies that, whether in whole or in part, could be modified according to changing economies of spares or the material conditions of roads."[103]

In a broad spectrum of activities, urbanites came to both prop up public infrastructure and prey on moments of failure.[104] If they could find no oppor-tunism in waiting, they simply suffered the cost in time and money of trying to get to work as wages stagnated and collapsed. Real wages fell 83 percent between 1974 and 1988 and men and women responded by abandoning for-mal employment altogether or subsidizing their wages with other endeavors that frequently emerged because of infrastructural failure. As I will discuss in the next chapter, this period was also marked by the proliferation of every-day forms of corruption and hoarding—or what the state called "economic sabotage." In many instances, corruption explicitly emerged because of the waiting and failure that people came to expect. Applying for a plot of land, one could be forced to either wait years without a bribe or avoid this wait through a well-placed payment. But more than just giving shape to new ways of making a living in an otherwise planned economy, waiting also became embedded within landscapes. Many urbanites took up jobs hawking food or selling charcoal. In these jobs, they both lessened their dependency on reliable transportation and brought neighborhoods goods that avoided disruptions of distribution due to long dwell times or lack of fuel. This was a new configuration of both urban time and space, that looked, in the eyes of the state, a lot like "loitering."

Waiting emerged as a publicly circulating metonym for failure or the nervous specter of failure. Standing still was either an act of resignation or an agonizing embodiment of inertia at a time when progress and economic sovereignty were predicated on "catching up." Of course, the dual questions that animated so many complaints written to the newspaper were, whose failure was this and who, consequently, was made to wait? And while waiting suggested stasis, it also had the pesky ability to spread: it was an animating force sometimes seemingly of its own making, stalling out those it struck in its recursiveness.

Landscapes of Deferral

In this final section, I want to explore how these ephemeral acts of waiting not only articulated new economies and itineraries in the city but fundamentally shaped Dar's built environment. The breakdown of Dar's transportation system both sprang from and created landscapes of deferral across the city. I take this term *deferral* from Jacques Derrida by way of the environmental philosopher Stephen Vogel. In the late 1960s Derrida coined the neologism *différance*, to identify the moment of transformation that happens when written text becomes enunciated speech. Différance captures the inevitable gaps between perception, speech, and reception: the unavoidable slippage between signs and their referent. In particular, the term plays with the fact that the French verb *différer* has two separate meanings: "to defer" and "to differ."[105] Vogel takes this term and cleverly applies it to built environments that, like speech, also exist in this articulated gap between intention and execution, and again between execution and their unfolding quotidian use. In any built landscape, there is no escaping an inevitable "moment of waiting . . . for the thought is never identical to the deed, and the deed itself never identical to what it does." Waiting becomes the temporal gap in which all built environments continually unfold, even if it is simply the space between what purpose a building was constructed to serve and how it is eventually reimagined and used by the city's residents. Essentially, nothing ends up being what it was first designed to be.

Focusing on how deferral has transformed landscapes offers a way to consider how both nature and communities shaped the city's roads long after their initial construction. Without the funds for maintenance, Dar's roads—paved and unpaved—were left vulnerable to the city's environment and climate. Dar's soil, topography, and watery landscape frequently caused the suspension of buses and cars but these landscapes were also in turn reshaped by waiting and breakdown. Nature could act in powerful ways, remaking streets into obstacles rather than the frictionless infrastructures of mobility they were intended to be. For example, the road map of the city could change in an afternoon. In a matter of hours during rainy season, standing water could reclaim and rework roads into impassable pathways, rerouting the circuits of the city daily, seasonally, and even permanently. One journalist reported after the beginning of the rainy season one year that roads in the neighborhood of Chang'ombe were under twelve inches of water and commuters waded through knee-high puddles hoping to find a bus at the next available stop.[106] The networks of creeks running through the city could also join forces in the rainy season, widening their paths to the sea and creating *mabwawa*, or ponds of water that sometimes remained year-round and could be deceptively deep.[107] Uncleaned and clogged storm drains also left water and debris sitting on the roads with no place to go. Dar's soil was a mix of sand and clay prone to erosion, leaving wide gullies under continued use as rains washed away topsoil. If unrepaired, this erosion would only be exacerbated during dry seasons. Paved roads could develop their own craterlike potholes that were most often fixed by leveling the bitumen entirely to the depth of the hole, creating depressions in the road where water could also collect and cripple unsuspecting buses.[108]

When driving on these transformed roads, their relative quality and maintenance determined how much fuel would be used on any given trip, how many passengers would be waiting, and how frequently a vehicle might break down. And after decentralization cut urban funding, only one or two technical personnel remained in the city's engineering department to fix roads for the entire Dar es Salaam region.[109] Persistently poor roads could lead the UDA to cut off services to certain neighborhoods during seasons when roads were "reduced to a cattle path" by torrential rains.[110] Without any drainage, both the craters and the accompanying pools of water could harm buses, sending them back to the UDA's repair garage to await spare parts.

Deferral lived in the moment when a footpath became a road, a road became a river, or a bus stop became a market. Nature has an arsenal of ways to make a city wait, and communities have their own strategies of reclaiming infrastructure back from nature in new forms. Urban residents in these neighborhoods fashioned palliative, short-term fixes to combat climate, rain, erosion, and waning bus services while they deferred to a future where there would be better roads and more reliable service. And since waiting for the bus inevitably produced walkers, it was pedestrians who reshaped the city's streets. In 1974, lines in Mwananyamala for buses were reported to be three hundred people long, snaking around corners and taking up the sides of roads.[111] And even after the legalization of dala dalas in 1985, 21 percent of the commuters who arrived at bus stops between six and nine a.m. were still left behind every morning.[112]

Walking is a category so overlooked in the modern city that it frequently does not exist in urban master plans, and rarely in newspaper narratives. We forget that walkers shape cities too. As the predominant form of movement across the city during this period, walkers still occupied the physical and metaphorical margins of Dar's roads.[113] But over time, pedestrians compensated for delayed and nonexistent buses by reengineering spaces that would otherwise serve automobility. The relationship of walking to a city landscape also echoes the distinction between written text and speech: it creates a "space of enunciation" where the planned becomes the provisional and the ideal is reshaped through the daily motions of the real.[114] This 1970s city was shaped by walkers' bodies that "follow the thicks and thins of an urban 'text' they write without being able to read it."

Walking in urban Africa can be more like "zigzagging, moving off and on the sidewalk or roadway," claiming space among cars, bicycles, buses, pushcarts, and hawkers.[115] And when there were no shoulders on the side of roads, people claimed ditches rather than waiting in vain for buses.[116] Residents of Mabibo, Mburahati, Luhanga, and Kigogo were described in the *Daily News* as taking up "walking-cum-jogging exercises" every morning "not out of the need to keep their bodies in good physical condition but out of transport difficulties."[117] This parade of joggers went from their homes to the next bus stop, creating "lengthy human chains" of commuters along the roads.

Beyond watching out for potholes, pedestrians also navigated and participated in the "colonization of the sidewalk" by vendors, "both itinerant and stationary."[118] Waiting for the bus produced its own economy that used the city sidewalks as its showcase. For example, after the Kigogo bus stop became a "white elephant," due to the infrequency of buses passing by, hawkers quickly converted the area into a weigh station to sell goods to commuters on their way to the next bus stop.[119] Acts of waiting—both anticipating the bus and finding creative and profitable ways to stave off the proliferating effects of shortage—became the fundamental force shaping the city's transport infrastructure.

This chapter has sought to bring together multiple forms of waiting and stasis in the city to consider how labor, landscapes, and infrastructure shaped one another. For a state that defined development around a philosophy of cooperative rural production, cities were suspect places where loiterers hid to avoid the hard work of nation building. Despite its focus on rural development, the fate of the Tanzanian economy was nonetheless contingent on an aggressive industrialization strategy throughout the 1970s. But workers were frustrated by the daily impediments they faced in navigating the city only to arrive at jobs that were increasingly irrelevant for helping them meet the needs of urban life in a time of shortages and shrinking wages. They were also frustrated to see their managers navigate a more privileged parallel urban landscape. From these two antagonistic perspectives, the city's fracturing infrastructure either was caused by or helped create the loitering classes who could not make it to work on time.

As formal labor became more irrelevant, the disruptions of roads and buses and the proliferating presence of walkers in the city offered up new opportunities outside of formal wage labor, further transforming the city's public spaces. But as M. A. Beinefeld argued as early as 1970, "so-called unemployment" in Dar "cannot be equated with free time and generally involves either looking for work or at least being ready to grasp an opportunity should one present itself."[120] Particularly as certain infrastructures broke down, supposedly *free time* became consumed by myriad forms of waiting: waiting for buses, waiting in line for food rations, waiting for the ability to buy goods, waiting for building materials, and waiting for employment.

In 1976, someone by the name of L. Joel wrote a letter to the *Daily News* titled "Like a Devil on a Mountain" that ruminated playfully on the cascading effects of the city's fractured infrastructure as well as the dangers of lateness for the new nation. The long letter narrates a moralizing tale of how the city's four-sided clock on the bustling Samora Avenue in downtown Dar was ruining lives. In need of repair, the clock showed a different time on each of its four facades: "Last week a clerk who regulated his punctuality by it got the sack after a number of reprimands. He had always trusted the clock so much that when he was told that he was late he was puzzled. Being a clock of the people he believed it could never be wrong. He used the northern face."[121] Continuing for several paragraphs, the letter narrates a pair of lovers missing a movie and ending their relationship, friends getting in a fistfight, a young man caught out after dark and robbed, and a tourist missing his flight to Nairobi—all the result of trusting various faces of the unsynchronized clock. In the following issue, the *Daily News* printed a photo of the offending clock confirming its state of disrepair. The clock was promptly fixed. In the ongoing struggle to shape the narrative of who or what was causing the suspended animation of labor and life in the city, perhaps this felt like a small victory for residents who were anxious to connect their inability to do hard work to the landscape and infrastructures of the city itself.

These daily acts of waiting sat in uneasy tension with much larger moments of deferral that were implicit in Nyerere's command for Tanzanians to "catch up." The very notion of needing to run while others walked was a plea for citizens to defer comfort, contentment, or relaxation until the future arrived. If the moment of independence was a moment of arrival for many

4.5. Newspaper caption: "'Devil on the Mountain' was yesterday removed from the clock tower by workers of the Comworks Maintenance Division in Dar es Salaam. It will be replaced by a new clock." *Daily News*, December 29, 1976.

Africans after a period of excruciating anticipation, liberation also quickly became a rhetorical device to justify more waiting. Calling on the memory of independence, states asked citizens to sacrifice personal rewards in the name of nation building: to simultaneously move quickly and wait for their own recompense.[122] This deferral of expectations expanded into the murky middle ground of the 1970s where decolonization existed as both a pivotal moment in time that had been achieved and an ongoing process that threatened to be ever left undone.[123] The fruits of independence seemed to be perennially waiting just on the other side.

Chapter 5

WASTING AND WANTING

Without eggs there will never be socialism.... Without the availability of
essentials, socialism will be stinking and the best thing to do then is to
throw into the sea whoever brought the policy of socialism.

Julius Nyerere

In 1985, Seif Sharif Hamad, head of the Economic Affairs and Planning Department for TANU, announced that the city council would take over control of Dar's valleys. After the city's unemployed youth were organized into
cooperative groups to farm them, the valleys would "flood city markets with
green vegetables and root-foods."[1] For officials, this project killed at least two
birds with one stone. It would help allay the city's food shortages and put
Dar's jobless to work. As a plan, it was nothing particularly new. There had
been any number of similarly coercive campaigns in the colonial and postcolonial era to compel the city's unemployed to labor productively in agriculture. It also echoed the state's ongoing desire to lessen inequalities between
town and country: if the jobless refused to leave the city, the country would
come to them. At a time of ongoing shortages, this was one of many efforts to
make the city provide for itself.

And yet despite Hamad's plan being a variation on a common theme, the
announcement garnered considerable attention in the following months as
concerned citizens wrote into the *Daily News* urging officials to reconsider.
These letters were not critical of the proposed work but instead described
the city's valleys as frighteningly polluted by both industrial and household
waste as well as local overflowing cesspits.[2] The resulting produce would
contain "a high proportion of the heavy metals cadmium and lead," wrote
in one concerned citizen.[3] Another letter writer was alarmed that Msimbazi Valley "receives untreated effluents from industries along Pugu Road
from Ubungo area ... one of the most heavily polluted. Not only industrial
wastes but also domestic effluents are dumped into these valleys. I know
that some gardens are being maintained in some of these valleys, but I highly doubt the hygienic condition of whatever is produced from these gardens
for human consumption."[4] The city's landscapes were serving multiple pur-

poses with a new intensity and people were beginning to worry about the consequences.

Dar's spaces betwixt and between formal definitions and designations had historically always frustrated the desires of the city's planners. Since the first master plan in 1949, city officials have sought to untangle the topography of the city's "Indian *dukas* [shops] and dwellings, mingling with the African and Arab huts" that "defied all efforts of municipal administration and sanitation" as well as to reform the equally transgressive spaces of its subterranean counterpart: the sewage system.[5] But neither the first master plan nor subsequent ones exerted much control over the city's heterogeneous spaces above or below ground; communities used these spaces for a variety of needs while municipal authorities perennially put off upgrades to the sewage system.

Planners had hoped that these very valleys transecting the city would separate residential areas and demarcate different urban zones. These waterways and their sloping banks served in the imaginations of colonial planners as ordering elements of the city's landscape. Dar's topography of slight rolling hills, intermittent waterways, beaches, and inlets created planning obstacles but also helpful natural boundaries. Marking the borders of neighborhoods, these valleys served on paper as cordons sanitaires and aesthetic buffer zones between the city's racially segregated communities.[6] Dar's African residents, however, commandeered this open space as a subsidy for low wages and high food prices. Indeed, communities living in Vingunguti and Hananasif settlements had long ago claimed the edges of Msimbazi for growing mchicha (a leafy green, similar to spinach), not far from where companies in the Pugu Road Industrial area discharged their waste.[7] Thus the layered use of urban environments now promoted by the city was not new to Dar. What made Hamad's project different was that Dar itself had changed in the meantime. First, it was not just communities but now city officials who were openly encouraging the muddying of old urban boundaries: crops and farm animals were now a common sight in the city. And yet, amid the call to claim Dar's valleys for agriculture, city officials also continued to close down informal markets and food stalls in the name of restoring urban order and stopping the spread of disease. What counted as dirt or disorder to city officials? These distinctions frequently had more to do with the state's conception of productivity than with public health or urban aesthetics.

This chapter examines the changing flows of goods and waste in the decade leading up to the city council's attempt to claim Dar's polluted valleys for agriculture. In the intervening years, the city had become a palimpsest of gray areas where the old infrastructures of both waste (sewerage and waste disposal) and want (formal channels for food and consumer goods) were fracturing, spilling over, and frequently remade entirely by communities, businesses, and the state. Examining how urbanites provisioned food and dealt with waste in a decade punctuated by profound shortages and eroding municipal services pushes up against how historians tend to narrate the growth of cities. Historians have generally written that the spatial and demographic expansion of cities is inseparable from the growing magnitude of commodity flows: bigger cities require more food and more goods, and thus create more waste. In the history of the industrial cities of the West, these widening material channels were then rendered invisible and commoditized: human waste removal and water delivery moved underground, food supplies were enclosed in railway cars and rubbish was carried away in trucks. Likewise, historians have pointed out that these processes of enclosure were part of a larger alienation of urban life from first nature (the notion of nature unmediated by human labor).[8] The concealment of these flows into cities and households (second nature) was part of a historic moment when the modern city came to embody the antagonistic opposite of the countryside in Western culture, despite remaining deeply enmeshed materially.[9]

In the case of Dar, however, the city's intensive growth spurt during the 1970s occurred at a time when many of its formal channels for goods were shrinking, sometimes completely drying up. This happened due to the collapse of the city's food supply. Tanzania went from being nearly food-independent a decade before to relying heavily on foreign food shipments and aid following an agricultural crisis that began in 1973 and lasted until 1976. Many factors contributed to the crisis, but three stand out as particularly significant. The announcement of compulsory villagization in 1973 led to the massive relocation of rural Tanzanians and the interruption of the growing season. Even those relocated only a few miles to new villages had their crops interrupted, and it took families and communities considerable time to reestablish their farming elsewhere. Second, a Sahelian drought that reached down into Tanzania also left fields dry.[10] While Ethiopia became the focus of international concern, a 1974 report by the Food and Agricultural Organisation warned that "Tanzania was on the brink of starvation."[11] Lastly, some of Tanzania's major grain producers, particularly in the Arusha region where 90 percent of wheat was grown, were anticipating nationalization of their farms and had consequently limited production, exacerbating shortages.

Compounding these factors, all rural producers were also transitioning from operating their own agricultural cooperatives where they would pool crops and negotiate pricing to instead dealing solely with the massive new parastatal the National Milling Company (NMC). In 1968, the NMC took over producer cooperatives and ultimately outlawed them. Farmers by law had to sell all of their cereal crops (corn, wheat, millet, sorghum, and rice) to the NMC, which set up buying stations across the country and introduced

universal pricing, irrespective of region.[12] They also instated year-round prices for grains, which left farmers with no incentive to hold back any of their
crops from the market once they were harvested: the fledgling parastatal was
flooded a few months a year with crops to buy, process, and sell while the
rest of the year they had relatively little.[13] Faced with rising oil prices along
with dwindling production and drought, the NMC struggled with its new
responsibilities for coordinating all aspects of acquiring and moving the nation's cereal crops to market. The NMC's low producer prices also encouraged farmers who could look elsewhere to seek out different markets. In some
cases that meant crossing the border to sell in Kenya or keeping cereal crops
outside of the NMC and selling them through informal market networks.
The result was a widely reverberating disruption in Tanzania's food supplies
even while export commodities were not as intensely affected.

By 1974, domestic food production was so problematically low that
the government was forced to spend its entire foreign exchange budget on
purchasing food overseas.[14] Urban food prices also rose, even as the state
sought to cushion the blow. The state's depleted treasury meant that salaries sometimes went unpaid and debts to local suppliers were unresolved.
Between 1974 and 1975, by the state's own estimation, the cost of living for
"middle-class servants" in Dar increased by 66 percent.[15] As fewer agricultural commodities made it to domestic and foreign markets, Tanzania's
available foreign exchange dwindled, pinching the budgets of the state's
emerging industrial sector of cement plants, textile mills, breweries, and
meat-packing facilities.[16] These factories relied on the importation of raw
materials that could only be bought with foreign exchange.[17] Thus, the collapse of agriculture had deep implications for the production and availability
of nearly all commercial goods as well as the availability of municipal funds
to deal with waste removal and improving critical infrastructures such as the
sewage system.

Following a brief respite in weather, export prices, and agricultural
production, Tanzania's war with Idi Amin in 1978 led to another period of
chronic shortages. Along with a shrinking budget and another peak in oil
prices, this brought on what was characterized as an acute urban crisis in the
early 1980s. Instead of the rise of satellite towns flush with agricultural and
industrial activity that the state had envisioned, Dar es Salaam was now five
times larger than Tanzania's next largest city and facing a period of dramatic and extended shortages just as more migrants arrived from rural areas.[18]
TANU's depiction of Dar as a parasite of progress and development only
grew more acute and unsubtle as they sought to discourage newcomers
from abandoning rural production. By now, the city was on the verge of infrastructural collapse and a dramatic metabolic rift had occurred. Regional
campaigns expelling urbanites and promoting urban agriculture sought to

transform the city's landscapes and populations into producers rather than consumers.

Thus, while urban histories usually narrate the alienation of city dwellers from rural landscapes and labor as the essential provisions of urban life are commodified and enclosed within urban infrastructures, the opposite became true in Dar. The paths of the city's material flows were literally left open to contestation, reclamation, and intervention. Communities, corporations, and state parastatals were soon entrenched in the daily provisioning of the city, reconsidering how goods could be made and what waste could be salvaged. These new channels became materially and discursively intertwined rather than isolated and enclosed. Urbanites and regional officials dramatically reshaped city environments through the imperative of rethinking resources under the threat of chronic and diffuse shortages.

In highlighting the remaking of materials flows, this chapter pivots between chronicling the reconstitution and alternative distribution of staple goods on the one hand, and on the other hand examining what happens politically as well as environmentally when food or waste spills outside its old channels. Forming clogs in fractured infrastructure or springing forth from points of leakage to forge alternate pathways of what "should be" invisible, both resources and "dirt" gave shape to a new contested geography of an expanding and descaled city. The need to reconstitute and rethink scarce resources and infrastructures also sparked ideologically charged conversations about what technology was best for a Third World country short on expertise as well as foreign exchange. Michael Lofchie points out that the two years of grain shortage beginning in 1973 ended up being "considerably more far-reaching than it appeared to be at first glance. Because of the primary importance of agriculture to the nation's economy, faltering food grain production quickly enmeshed the entire society in a complex series of deeply interrelated political and economic crises."[19] It also fundamentally changed Dar's urban landscape. What was productive and what was polluting became a matter of debate and depended on one's vantage point of the city and economy. Cleaning up the city and cleaning up the economy became state projects of restoring legibility in order to save socialism from economic "saboteurs." Urban communities, meanwhile, searched for their own forms of legibility and security by which to get by in tough times.

Wanting

When formal channels for food supply began to collapse in 1973, Tanzanians first sought relief through their kinship and community ties. Families reanimated old connections and made new ones, but these new channels did not always flow in the same direction they had previously.[20] In many rural areas "the direction of economic change was reversed; it was a matter no longer of moving from a subsistence lifestyle toward widening participation in the marketplace but of moving from a mixed economic pattern back toward subsistence cultivation as a strategy for economic survival."[21] But this reliance on subsistence practices was never an act of closure to the outside world nor a wholesale retreat to some former way of life. As Bryceson writes in her important work on urban food supply, a family member in the city was crucially important to this new subsistence strategy; they "received innumerable obligatory requests, if not dictates, from . . . up-country extended family, which could entail heavy social costs if ignored."[22] Urban family members often found themselves caring for rural relatives coming to look for jobs or access to urban services, subsidizing their economic needs until they got on their feet. Food shortages made the advantages and importance of the city all the more urgent for rural Tanzanians, and Dar remained a crucial place for families to inhabit, as part of their attempt to diversify the means of their survival and to attend more broadly to the acute absences in their lives. As one *Daily News* reporter wrote in 1973 after visiting his rural family, they were full of expectation of what a visitor from Dar might bring with them:

> When I reached my small village twenty-five miles from Masaai on a short feeder road, everyone was delighted to see me. I thought all of it was part of my family spirit and good neighbourly cheer. But no, quite a good deal of it was the result of calculated relief. They saw in the "son" from Dar es Salaam the answer to their shortages. Where is grandfather's blanket? Where are my khanga? And what about Kaniki (women work clothes, equivalent of the Khanga) we have not seen them here for six months! Have you brought any sugar with you? Bright and early the next morning I was at the local duka. There was not much I could get from the local duka. The sugar had got finished the week before and the shopkeeper did not think he would have any luck at the not-so-local duka when he went to see the supply position that morning as he was intending to do. The word "shop" is an awful misnomer for the bare shelves. He "sells" maize, flour, beans, sugar (if they get it), cigarettes, matches, tinned milk, tea leaves, female decorative trinkets, writing paper, and packets of FAN powder soap.[23]

But increasingly, this exchange of goods went both ways. On the way back to the city, families passed on to their "urban son" bags of maize, rice, coconuts, or cashews to sell in unofficial markets or simply to live on for a short while until more could be found.

These strategies, however, ran counter to the state's desire to minimize the movement of people and goods outside of official channels and stem the retreat to subsistence agriculture. In rural areas, Nyerere reinstated old colonial bylaws requiring families to cultivate a minimum acreage per family. In some regions, before farmers could board a bus or train, attend a cattle auction, or go to a beer hall or ceremonial dances, they first had to present a certificate from the local TANU/CCM office to prove they had planted a required minimum number of acres.[24] The government also set up roadblocks to check for any unauthorized commerce, but patrols sometimes turned a blind eye to peasants traveling to the city with fewer than five bags of produce.[25] Soon, being stopped and searched for "contraband" food became "one of the most common experiences in the daily life of rural Tanzanians."[26] Rather than stopping the transit of goods within family networks, this likely became a reliable source of bribes for Tanzanian officials as they also sought new strategies of accumulation.

Meanwhile, the state response to urban food shortages was generally twofold: to control cereal supplies by dispensing staple foods through state shops, and to urge urbanites to produce their own food. Both securing food supplies at state cooperatives and growing one's own food became incredibly time-consuming tasks. The ebb and flow of official supplies meant each family had to develop a daily strategy for trying to beat the unpredictable odds of urban distribution. Finding staples like rice or *sembe* (maize) could entail first waiting in line at a state-run shop, if it was currently supplied. Families could only get food if they first registered for a card from their local ten cell unit. They could then take this card to local state-run shops when new supplies arrived. Knowing a delivery was arriving the next day at a particular shop could mean getting up in the middle of the night to get a spot in line before supplies were depleted. Others attentively looked out for supply trucks around the city and ran after them to their destination.[27] One woman I interviewed remembered following a supply truck full of cassava flour to its destination hoping to secure a good spot in line, only to watch the person in front of her receive the last bag. But she also recalled the woman turning around to give her a kilo because she had a baby.[28]

If a family could afford it, they could alternatively rely on getting goods without waiting in line. In this case, they sought out "queueing boys," young men who made a living by standing in line to buy goods at official prices, reselling them at market rates.[29] These young men represent one way that waiting itself was fashioned into an occupation. For those without jobs, the very ability to wait could net a small margin for survival. Yet queueing boys also embodied Nyerere's main complaint of urban life: that young migrants to the city were not producers but rather an ever-expanding population surviving off of a repertoire of margin-making activities.

While one family member waited in line for staples, someone else might head to the fruit and vegetable markets. Radio Tanzania's weekly broadcasts announced the price of produce and vendors recalled savvy customers coming in armed with this knowledge to avoid being overcharged. While all fruits and vegetables were supposed to be routed through the main auction house in Ilala to then be sold by smaller vendors and markets, many sellers sourced their goods from families, neighbors, or local producers who also hoped to sidestep official market prices.[30] Acquiring produce was easier in more central areas of the city, leaving those on the periphery to rely on buses, bicycles, or walking to access them.[31] Yet in time, stranded customers frequently became small-time vendors themselves, providing access to fish, dairy, vegetables, and fruits to the city's expanding outer neighborhoods, particularly as public transportation faltered.

Local markets also doubled as places to gather information or gossip about where other supplies might be found. Sellers known as *walanguzi* (sing. *mlanguzi*) operated their own "gray market" within these spaces. While these illegal streams of food and consumer goods were essential, they also produced frustratingly vulnerable transactions that left many feeling cheated by high prices or scammed into buying faulty goods. One woman recalled making trips to Tandika market where she would look for small piles of rice or maize flour on tables as a symbol that someone was selling these items: "You just go there and put money on the table and leave the place. The walanguzi took the money and brought to you rice if it's rice."[32] There was no way to double-check that a kilo was indeed a kilo or even to make sure you got what you came for. Another interviewee recalled watching a woman buy a shirt from a mlanguzi who told her she could not open the package for fear the police would see their illegal exchange of goods. Once she was able to open it, she realized the shirt was nothing more than a sisal sack.[33] To evade the police, other walanguzi traveled from house to house to sell their wares. After checking for police, they would enter buyers' homes to sell *khanga* (cloth), soap, shirts, or other hard-to-find goods. Sometimes, wearing wide-legged jeans, the sellers would wrap the contraband cloth around their legs underneath, revealing it only once inside the house.

While regional authorities policed illegal channels for buying and selling goods, they also urged urbanites to cultivate their own alternative flows of food, even mandating urban agriculture. Urging Dar's residents to start farming wherever they could, whether it be their front yards or in peri-urban areas, Nyerere claimed that a worker who had "grown his own vegetables or paddy" was someone who could meet "the increased cost of sugar or *khanga* with little less difficulty."[34] Rather than rerouting urbanites from formal markets like buying illegal cloth did, growing one's own food ostensibly left families with money for purchasing other goods.

At least initially, the city's main cultivators were frequently not the male worker invoked in Nyerere's speech but women who cultivated supplemental gardens near homes and in the urban margins to avoid both expulsion and hunger in the city.[35] Living both in Dar and its peri-urban zone, these women sometimes walked miles each way on weekends to attend to both worlds.[36] As one man I interviewed recalled, his wife supplemented his factory income during Dar's food shortages by farming in Kimbiji, about thirty kilometers from the city center. In total, she spent about six years tending to a small farm "during the time when food was obtained by a list of names."[37] Ultimately, it was this money that allowed them to buy a plot of land in Mbagala.

Wage-earning women also frequently ran informal businesses on the side, selling prepared foods, produce, or products they could make by gathering local materials such as *makuti* mats woven out of palm leaves.[38] Certain factories such as the Friendship Textile Mill also sponsored classes for learning these trades. Usually the profit from these sales provided a modest buffer to family income but it could provide a much more profound sense of security and autonomy for women.[39] One woman proudly claimed when interviewed in the 1970s that she made more selling beans, rice, *maandazi* (donuts), and tea than her husband's wages, and as a result, "whether he [her husband] is at home or with another woman outside, I don't care. I have my own money and the house is mine."[40] In this way, the peri-urban environment could foster women's economic autonomy. In 1974, just one year after Tanzania was hit by food shortages, 70 percent of homeowners had vegetable gardens and half had fruit trees, and by the 1980s even renters were farming.[41] The 1976 Household Budget Survey of Dar es Salaam reported that half of household food consumption in the city came from farms unmediated by market exchange: the poorer the family, often the greater the percentage.[42] This percentage only grew over the next decade as Dar's food channels continued to be dramatically reconfigured. And as the urban crisis worsened in the 1980s, men, too, abandoned their jobs to farm and run small businesses.[43]

Urban agriculture has long been a central facet of how families "make do" in African cities, but what may be remarkable about Dar in the 1970s and 1980s is to what extent these practices were encouraged by the state. In fact, cultivating a small *shamba* was not only allowed but became the way to secure tenure in the city if you were otherwise unemployed: cultivating land in the peri-urban zone allowed urbanites to procure a shamba card that they could show the police as a way to avoid expulsion. Beyond families, Nyerere also required parastatal industries and private businesses to contribute to the city's food supply. In 1974–1975 the president announced Operation Kilimo cha Kufa na Kupona (Operation Farm for Life or Death), which required Dar's "textile factories, breweries, cigarette and soft drink plants, shoe factories, and other parastatal institutions such as TANESCO [the electrical company] and the University of Dar es Salaam" to donate time, manpower, and resources in the form of transport vehicles, tools, and money to produce food crops in assigned plots on the outskirts of the city.[44]

When touring these emerging agricultural operations—including farms started by eighteen companies and institutions in the city—Nyerere noted that these new ventures and village were "lucky because they have a ready market."[45] In theory, these companies were to sell their crop yields to the NMC, but it is unlikely that this reliably happened.[46] More likely, Kilimo cha Kufa na Kupona funneled precious state parastatal funds in the form of employees' wages, time, and state resources into developing subsistence food channels for workers' families as well as a small margin of extra income. By 1976, 108 parastatal enterprises were farming, some at huge losses "in terms of capital and personnel."[47] In 1985 the general manager of the Tanzanian Cigarette Company, Brown Nwilulupi, urged an audience full of employees of the National Development Corporation to each farm for themselves since "an enterprising Dar es Salaam rural dweller could earn about 8000 a month out of mchicha from a small garden."[48] Nwilulupi shared his own experience: "I grow vegetables and oranges. I also have a cow. In fact, I left the public house in Oyster Bay long time ago before I retired to live in rural Dar es Salaam. Do not fear going into the suburbs. Life there is cheaper than you imagine. I have no pension after serving the nation for 35 years, yet I can manage to live on what I get from my farm." The future of the city lay increasingly in urban residents going to ground in the periphery. By 1985 the state Prison Department had also created a 1,200-acre farm of cassava and coconuts tended to by 150 inmates in the Wazo Hill area. That same year, the director of prisons asked for more land to expand their farm to 40,000 acres, including room for beef and dairy farming.[49]

These farms created as an emergency stopgap effort for food and raw materials soon became part of the permanent fabric of the city.[50] By 1979 there were an estimated forty-two agricultural and fishing villages established in the region that were as reliant on selling their goods to the city as the city had become on these new producers. Residents of these villages—which were home to eighty-three thousand people—walked headloads of food into town to sell.[51] When smaller feeder roads were washed away in the rainy season or transport was scarce, the city could be cut off from these important food supplies, since the city's stock of perishable fruits and vegetables lasted only a day or two. This caused prices to skyrocket for urban shoppers whenever roads deteriorated. During harvests, staple food crops were often in competition with each other for access to buses and trucks. Banana sellers were forsaken for orange sellers during the orange season if drivers could negotiate better rates with them.[52] Shortages of roads and vehicles left urbanites vulnerable to high prices for fresh goods and further encouraged having one's own land.

By expelling city dwellers into peri-urban production and mandating parastatals to farm, both the state and citizens transformed Dar's frontier into the city's local food supply. Far beyond the initial vision of a worker supplementing his income with a small plot of land, by 1983 any industry in the city "utilizing agricultural raw materials" for production was required to establish peri-urban plantations to meet at least part of their material demands.[53] What had first been imagined as a way to ease urban dependencies on erratic food supplies expanded into an effort to keep factories online that were foundering from raw materials shortages. By this point, what constituted rural and urban space was no longer straightforward in practice, even if a planning boundary still hypothetically dictated what was "city" and what was "country." Factories, workers, and families all transgressed these boundaries to survive life in the city.

Perhaps those who most notably transgressed and challenged these lines messily were animals. The city council conducted an animal census in 1982 and found that Dar was home to "4 donkeys, 53 horses, 138 sheep, 176 rabbits, 286 turkeys, 1,343 head of cattle, 1,418 goats, 6,508 pigs, 7,032 ducks and 424,028 chickens."[54] University professors were nicknamed 'the banana and chicken petty bourgeoisie' because of the common practice of supplementing their salaries with small poultry and farm operations, sometimes out of town but otherwise on the university campus.[55] Several university residents bought and raised dairy cows on the hilly green campus. Starting out with a few chickens and four dairy cows, one professor ended up with fifteen scrawny heifers to supplement his income. In fact, he said, the university sold their tractor (for which they likely lacked fuel) and just let the cows graze instead. In announcing the results of the animal census Dar's city director, David Mgwassa, chastised residents for forgetting that "they were not allowed to keep animals in residential areas." But clearly the regional and state authorities had also actively blurred the line between city and country, legal and illegal, and most notably, between what was productive and what was polluting.

Unsurprisingly, these agricultural enterprises, livestock, and the profusion of markets made Dar a messier city. Staving off want was leading to more dirt: the new emerging flows and channels of goods traversing the city sprang all sorts of leaks, some intentional and others collateral. Sometimes these were harmful and other times they were generative. Indeed, what counted as dirt frequently depended on one's perspective. A pile of organic waste was also food for the city's livestock. These liminal materials and spaces that were propping up the city came to exist in the volatile temporalities of urban governance that could swiftly swing from long periods of municipal neglect or forbearance to intense periods of scrutiny and eradication. For example, when the municipal government in 1979 announced that they would clean the overgrown areas around Dar's roads to give the city a "hygienic face-lift" many residents feared the state were "thrushing away people's 'self-reliance' farms in the city" after having demanded five years ago that they plant "every inch of their backyards and front gardens for food plants."[56] State and regional demands for agricultural production made for increasingly dirty demarcations of what constituted proper use of urban space as well as illicit enterprises. Alongside the growing presence of actual dirt, the party also developed in the press and through public speeches a vibrant discourse on waste.[57] Through public campaigns they took aim at "economic saboteurs" and street hawkers by invoking the need to discipline and clean up Tanzanian society along with its city streets. While growing food was encouraged, selling it along the street was not, and sanitation became the justification to curb such practices. Invoking sanitation and selectively pointing out and making visible certain forms of dirt legitimized demands for reining in new markets and modes of transactions when the state desired.

Blurred Lines

Dar's sanitary boundaries were also further muddied due to the foundering of municipal services. The budget for city sanitation services was dramatically affected by decentralization in 1973 and the dissolution of urban councils. Most households in the 1970s therefore had little opportunity to get rid of accumulated waste other than to bury it. In fact, sometimes waste was used as a form of landfill for those who took up residence in Dar's hilly areas, accumulating and filling in slopes and valleys over time. Dar's main waste dump, Tabata, had passed its capacity in the 1960s though it remained open and overflowing until 1991.[58] Efforts to develop an alternative new site stalled out when residents protested that it would attract a similar community of scavengers that made marginal livelihoods from reclaiming waste in Tabata. But even if a new dump had been opened, only a small fraction of the waste generated by the city would have ever made it there. Just a year after decentralization the city's fleet of thirty-three garbage trucks for waste removal had been whittled down to nine that were usable: "The rest are broken down and could not be repaired" due to "shortage of spare parts or because of the cost of repairing some of them would be almost equivalent to the cost of buying new ones."[59] The tonnage of waste picked up in the growing city meanwhile decreased from around 175 tons per day in 1968 to only 100 tons a decade later as the city continued to grow.[60]

Like the city bus, Dar's residents were at the mercy of a broken-down fleet that required expensive fixes with foreign parts. The *Daily News* reporter Felix Kaiza urged regional authorities to consider a new approach to cleaning market places, residential areas, hospitals, and hotels of Dar that look "no better than dumping grounds."[61] He was particularly frustrated that thinking about trash had always led municipal officials to import garbage trucks: "Every time we think of garbage collection our minds are taken to means that are thousands and thousands of kilometres away from us—means which we can get through bleeding our national economy white." Kaiza worried that the very acquisition of trucks from abroad with foreign exchange was eating into the budget for municipal maintenance needed to operate them. Instead,

he suggested that municipalities needed only to purchase much cheaper tractors with foreign exchange and then they could craft their own trailers with
local steel that could be hooked up to the back of a fleet of tractors and easily
placed around commercial areas and homes. The neighborhood of Kariakoo,
in fact, already had one with three locally made self-tipping trailers. Alternatively, water tanks could be attached to the tractors and taken to hospitals
and schools when the city's water went offline.

A year earlier, in another attempt to avoid reliance on the city's fleet of
garbage trucks, the district of Kinondoni introduced a fleet of donkeys and
carts for waste-removal services, noting that they would also "reduce petrol
consumption."[62] Temeke and Ilala districts also considered switching to animal power while first waiting to see how Kinondoni's animals adjusted to
the climate and traffic. Of course, the donkeys would discharge their own
waste while picking up the city's. Nyerere also urged citizens to keep the
town clean on their own without relying on garbage trucks. In 1983, starting
a nationwide campaign for urban cleanliness, he argued that "lack of foreign
currency" was not responsible for the "deteriorating situation of urban centers." After a three-hour meeting with the urban council, Nyerere reportedly
asked, "What type of foreign exchange was, for instance, required to fill up
a ditch on a muddy road left open for mosquitoes to breed in?" While aware
that the city was running on 9 trucks when it needed 110, Nyerere still saw
the city's trash problems as symptomatic of bad habits rather than a shortage
of foreign exchange.[63] This became a key tension of the 1980s as the economy
struggled: had self-help failed, or were Tanzanians at the limit of what they
could do to "help themselves" without better infrastructures and services?

Regional authorities also attempted to find solutions for Dar's waste that
avoided reliance on trucks and unavailable funds. They routinely inaugurated
campaigns against "filth" and "dirt" that framed Dar's waste problem as one
that could be solved through personal responsibility. Ten cell leaders ordered
residents to clean their homes and would afterward conduct inspections. Indeed, one of the most actively blurred lines of all was who bore the responsibility to keep Dar clean. Women were frequently framed as the sources of
dirt. "Some women tenants are so irresponsible that you will find all sorts
of dirty things, including worn-out baby napkins stuffed in their toilets,"
complained one delegate from the National Housing Corporation to explain
why government-owned housing was dirty.[64] "Country bumpkins" who had
recently moved to the city were blamed for not knowing how to properly deal
with the new forms of waste. Campaigns against dirt served a dual purpose
as both didactic and hygienic missions to clean the city and groom urban
housekeepers. Women's organizations were likewise tasked with taking up
the cause of cleaning up the city. The Dar es Salaam branch of the women's
organization Umoja wa Wanawake wa Tanzania (UWT) frequently execut-

ed these campaigns, taking to the streets in groups armed with brooms and dustpans.[65]

These campaigns could also be frustrating exercises in futility. Collecting one's domestic waste could only be part of the solution if it were picked up and taken to the dump. When it was instead left to rot in overflowing and scarce rubbish bins and redistributed by animals, many residents felt they were being asked to do the impossible task of making waste literally disappear at the moment when it was proliferating the most. "We city residents," one wrote to the *Daily News*, "are told to keep our environments clean and in fact we do that, but where are the dustbins and trucks which used to collect garbage? We know that there is a shortage of vehicles, but even once after two months could help to minimize the problem."[66]

Communicable diseases often catalyzed these daily struggles with waste into moments for municipal intervention. In these moments, certain populations became identified as the offending vectors. One of city's most acute moments of reckoning came in 1978 when Tanzania was hit by a nationwide cholera epidemic. In January, 238 people had already died and more than 3,000 cases had been reported. Dar residents were warned over the radio and in newspapers that cholera was spread through "people and dirt."[67] To contain the epidemic, the state shut down unpermitted restaurants, food market stalls, and sidewalk bars and canceled meetings and social gatherings in affected areas, pointing to the city's informal enterprises for spreading the disease. With cholera outbreaks on the rise, Hussein Shekilango, in the prime minister's office, announced the reinstatement of urban councils for the first time since decentralization, realizing that the city's ability to respond to such a threat was immobilized by its weak and fractured municipal governance.[68]

With few resources or institutions to fight the disease, the newly reinstated urban council continued to advocate for self-reliance and discipline to confront the spread, along with policing "dirty" spaces of commerce. Shekilango also urgently ordered all urban residents to "transform the urban areas into areas of production." This was likely a confusing directive for many, since the state had sought to arrest the spread of cholera precisely through targeting what for many were the productive spaces of the city: the vendors, shops, and *magenge* (street restaurants) that crucially supplied and distributed food to the neglected corners of the city. There was no easy delineation between productive enterprises and the waste they generated. With official channels for both food and waste in disrepair, these mixed spaces of production and consumption were frequently essential sources of food, money, and goods.

The *Vibrio cholerae* bacteria certainly did thrive in the leakages of a city of layered use and aging infrastructure; for years the disease has seasonally arrived with the rains in Dar. But it is more precisely spread not by dirt or people but in brackish water, usually showing up when untreated fecal matter enters water supplies. That water then may be used to wash produce at markets or be consumed as drinking water. These problems started below the city's surfaces as well as in its clogged drains and cesspits. By the late 1970s the sewerage system served only 12 percent of the city while another 10 percent had access to septic tanks.[69] The rest relied on pit latrines, and by the time of the cholera outbreak there were only sixty thousand latrines for a population nearing one million.[70] Pit latrines also relied on cesspit emptiers and, like the city's garbage trucks, a majority of the emptiers were stuck in garages waiting on spare parts.

By the 1980s the dire situation of Dar's sewer had attracted the attention of the World Bank, which noted in its report that since decentralization, "sewerage and sanitation operations and pit latrine emptying services suffered and virtually ceased."[71] Additionally, only one waste stabilization pond out of nine worked and only one of seventeen pumping stations for sewage were in operation, with many sewers completely blocked. Dar's subterranean environment was bursting at its seams and as a result around twenty-two thousand cubic meters of untreated sewage a day was jettisoned into the sea, with peak flows around fifty-six thousand cubic meters.[72] Dar's industrial wastes were also "substantially uncontrolled and untreated" with about thirty industries near Msimbazi Creek categorized by the World Bank as "serious sources of pollution," disposing their wastewater into the creek despite legislation to prohibit such practices.[73]

Dar's gray areas were becoming the city itself in both productive and harmful ways. The urgency of exploiting the urban environment—frequently at the behest of the state—collided and collaborated with the city's debilitated municipal sanitation forces and the unmonitored effluent of industries. The more multivalent any city space was, the more competing uses could jockey for what would be considered matter out of place. What could coexist for years of economic crisis as shared space became at points of sanitary crisis sharply defined and contested. And yet what seemed to determine whether a space was polluting had more to do with whether it was productive in the eyes of the state than whether it was toxic. Naming and seeing dirt was never just a sanitary act but a political one, employed to point out transgressive populations as well as unruly spaces.[74]

Reclamation and Reinvention

While leakages and gray areas could arrest urban productivity and become life-threatening, they could also be part of innovative new ways to steward materials and foster national resource sovereignty. The practicality of poor urbanites drove their own ethic of reuse and thrift, but their efforts also struck a chord with the state's call for self-reliance in both local and national realms and as part of a broader Third World push for resource independence.

For Nyerere, wastefulness seemed to reflect a new carelessness by Tanzanians too eager to adopt Western habits. In looking back, in 1977, on ten years of ujamaa, Nyerere seemed nostalgic for a time when less wastefulness and disorder seemingly abounded, particularly among bureaucrats. Instead of scorning bicycles in favor of cars and wasting paper with the excessive use of envelopes, recalled Nyerere, agricultural officers used to "walk from village to village on duty, spending the nights in people's houses or in tents. . . . . Serious attention must be given to every detail of expenditure, and the question asked 'How can the job be done more cheaply'?"[75] As in many socialist economies where shortages became a way of life, thrift became the responsibility of everyone and an issue of national identity.[76] But for most Tanzanians, still one of the poorest countries in the world, there was little need to moralize about waste. Fashioning something out of nearly nothing was an old skill of survival in urban Africa: migrants arriving in colonial cities for work had routinely turned to waste streams for resources that meager wages did not otherwise accommodate. It was a strategy that went hand in hand with the practicality of urban agriculture. They repaired and reinvented materials, extending their usefulness or creating new objects entirely. Navigating the new territory of both consumer objects and urban poverty, townsmen created charcoal stoves from car doors, lamps from oil tins, and tambourines from bottles.[77]

Writing about the stewardship of old objects in colonial America, Susan Strasser notes that in the face of material scarcity, "everyone was a *bricoleur*" working to reshape and reinvent otherwise worn-out and broken things. In postcolonial Dar, shortages similarly sparked attempts to develop new objects from old ones.[78] Reimagining the possibility of materials plucked from the waste stream happened across the city: in parastatal factories short on parts, in illicit garages fixing broken-down buses, and in Tabata, where scavengers salvaged durable materials from the municipal dump. Facing daily power outages, those living in houses with electricity turned to other sources of fuel. Hospitals were routinely short on even basic supplies, necessitating that patients bring their own in order to have a procedure, and "university instructors, lacking paper, wrote their syllabi on the chalkboard; their students took notes on the margins of scrap paper . . . even the most basic consumer goods such as batteries, tools, light bulbs, automobile parts, and kitchen utensils became difficult to obtain."[79]

As the ongoing reappraisal of space and material drove the refashioning of the city's green edges into agriculturally productive enterprises, it also transformed the heart of Dar's industrial sector. Places like Gerezani and Chang'ombe became increasingly dedicated to these practices while also becoming known for their growing effluent flows.[80] These neighborhoods were close to the train tracks, downtown, and the harbor, and have always been the locus of the city's industrial activity. By the mid-1970s Gerezani was brimming with vehicle mechanics, metal welders, plumbers, and small-scale industries making or reimagining everyday consumer goods.[81] One of the main organizations to thrive in these areas was the Dar es Salaam Small Industries Cooperative Society (DASICO),[82] which began as a cooperative society in 1967 with fifty members, mostly trained in industrial repair and manufacture under the British and Germans. The idea behind DASICO was to localize and collectivize artisans otherwise spread across the city's out-door spaces, where they faced being shut down by local authorities for operating informally and polluting the city.[83] Within this location, the informal were in effect claimed by the formal sector, even as many workers attempted to avoid taxes and facility fees.

Members of DASICO (nearly entirely men) were known as *mafundi chuma*, and they forged new objects out of old ones by sharing workspace at a "common facility workshop" and learning from one another.[84] Reworking wire, steel rods, old car doors, empty vegetable oil tins, and oil drums, these artisans bought their materials from men called *skrapas* (scrappers) who scavenged across the city, bringing items to the DASICO to sell.[85] Men and women sorting waste at Tabata dump provided 60 percent of the raw materials for DASICO and similar organizations, but the most valuable scrap material never made it to the dump.[86] By the mid-1980s the dump was home to a long-term community of scavengers. The oldest scavenger, Mr. Takataka (Mr. Garbage), had been scavenging for twenty years.[87] Mr. Takataka had witnessed a shift over the years from collecting at the dump to scavengers moving into the center of town for more valuable materials since waste no longer made its way out there.[88] The city council over time also grew mostly ambivalent to the practice of scavengers: "At first the council fenced off the dumping area, but as the fill grew higher the fence was buried in the waste. Since then the attitude of the city council has apparently been neither in favour of nor against scavenging."[89] Working from a copy or sample, the wafundi chuma would then shape their scavenged materials into a broad range of basic consumer goods. In the early days of operation, DASICO had a marketing officer who would sell the finished goods and charge a 7 percent fee that went to the maintenance of the workshop. By the 1980s, however, artisans found their own buyers or sought larger orders from businesses, or people would directly come to the workshop and attempt to make their own transactions in order to keep more money.[90]

After 1973, DASICO became part of a newly established parastatal known as the Small Industries Development Organization (SIDO), established to satisfy TANU's directive Agizo Juu ya Viwanda Vidogo Vidogo Nchini (Commission on Small Industries). TANU defined a small-scale industry as "any unit whose control is within the capability of the people, either individ-

5.1. SIDO workshop in Chang'ombe, Dar es Salaam. Photo courtesy of Tanzania Information Services.

ually or cooperatively, in terms of capital required or know-how, it includes handicrafts or any organized activity based on the division of labor."[91] Motivated by the breakdown in transportation networks throughout the country, and the desire to locate industrial centers outside of Dar es Salaam, the goal of SIDO was to bring a "technological revolution to the rural areas" by promoting the local production of clothing, hardware, agricultural byproducts, handicrafts, and building materials.[92] SIDO desired to create industries that would reduce national reliance on large-scale industrial production.

In addition to salvaging reusable materials, SIDO also sought to identify overlooked locally available raw materials. Just like the university's Building Research Unit promotion of alternative vernacular building materials, SIDO experimented with making familiar goods with new local resources. In fact, they attempted their own cement alternative out of lime and pozzolana (volcanic ash), both locally available raw materials. A SIDO team also worked out a method of turning agricultural waste into gas that could run diesel generators for operating grain mills. As this method was designed to run on corncobs, some worried that there might be a better use for this waste and feared it could also lead to a dependency of its own, promoting local deforestation for more corn production. SIDO also supported the establishment of five iron foundries throughout the country for making farm implements. Using sand from Kisarawe, Dar's foundry took scrap metal and

5.2. SIDO workshop in Chang'ombe, Dar es Salaam. Photo courtesy of Tanzania Information Services.

rejected hoes purchased in bulk from Ubungo Farm Implements (UFI) factory in Dar and reforged them to sell. In 1982, eighty tons of scrap from UFI was sent out to different blacksmith groups but these stockpiles of precious reusable materials seemed to have sat in the back of local SIDO offices, never distributed.[93]

With these and other endeavors, SIDO struggled to fulfill its goals of transforming and decentering industrial production. For starters, many of their enterprises did not sufficiently advertise their products to the public.[94] And despite TANU's directive to inaugurate a rural self-help technological revolution, SIDO products also faced competition from other parastatals who sold competing items produced by larger factories. In many cases, these projects also had to be started "from scratch," so they took far more bureaucracy and manpower despite their "small-scale" size. Even within the parastatal itself, several senior SIDO officers, when interviewed, said that they believed rural craft and urban informal sector units "weren't important." They felt instead that Tanzania needed "modern small- or medium-sized factories using advanced techniques."[95]

The new push for small industries was perhaps most successful in cities and towns where they were less engaged in specific projects and instead provided industrial common facility workshops where individuals made cooking utensils, farm implements, door clasps and bolts, spare car parts and animal feeders, wheelbarrows, cupboards, steel cabinets, windows, doors and chair frames, office furniture, beds and sofa sets, school desks, bread ovens, and ploughs.[96] Sixteen new industrial estates were created in different regions, each with a common workshop and in total eighty-nine different industrial units that could produce essential goods.[97] However, the electricity supply parastatal TANESCO was unwilling to supply some of SIDO's new industrial estates with electricity, limiting its diffusion.[98] Wafundi chuma could use the workshops for their own or joint enterprises, continuing to foster their skills and mentor a steady stream of apprentices. SIDO's survey of Dar's small-scale industries reported that there were 448 such industries in 1975 and just two years later 2,095 small-scale industries were listed.[99] Thus, SIDO's most important role was likely in giving space and communal resources to informal craftsmen who were themselves constantly considering the possibilities of reinvention.

Rethinking Technology

Focusing on the stewardship of objects went beyond reconsidering material flows in small industries and became a national effort to also rethink the adoption of technology writ large. In this section and in the next chapter, I explore different ways that "appropriate technology" was imagined and promoted both transnationally and within the Tanzanian state. Here, my focus is on how the state used the term to consider larger technologies on the scale of the factory or small-scale industrial manufacture that would be "relatively simple, sturdy and easy to operate and maintain."[100] In the next, I explore what is more frequently considered "appropriate technology": small, mobile, simple technologies that were imported through aid programs and designed for broad adoption at the household level.

When the Tanzanian economy first faltered in 1973, international development economists voiced concern that Tanzania did not have the resources to move forward with its ambitious plans for industrial development.[101] The state was still in the middle of its Second Five Year Plan (1969–1974), which called for an aggressive expansion of the nation's industrial capacity, hoping to foster Import Substitution Industrialization (ISI).[102] This was the general theory that one of the most crucial steps of decolonization for less-developed nations was to develop domestic production facilities that would allow them to stop importing basic manufactured goods from particularly the United States and Europe. This was a key step in shifting away from the colonial economic order that had relegated colonies to be the producers of raw materials and the captive consumers of foreign manufactured ones. Staying the course of ISI for an overwhelmingly rural country like Tanzania would require an immense transformation of its manufacturing sector. This was particularly true since neighboring Kenya had received far more industrial development during the colonial period, with the assumption of a common regional market shared among Tanzania, Uganda, and Kenya.

But even following the first oil shock (and the exhaustion of all foreign exchange funds in 1973 for food imports), Tanzania announced another comprehensive Basic Industry Strategy (BIS) in 1976. If anything, the unfolding oil crisis ratcheted up the state's desire to domesticate manufacturing and free up foreign exchange reserves otherwise spent on imported goods. While staying the course of rapid industrialization, with BIS the government began more aggressively reappraising what sort of technology was most appropriate for cultivating economic sovereignty in a rural, socialist nation. In some ways this had been a key part of Tanzania's industrial thinking from the beginning. In 1968 Nyerere noted, "When we built factories which serve the whole nation, we have to consider whether it is necessary for us to use the most modern machinery which exists in the world. We have to consider whether some older equipment which demands more labor, but labor which is less highly skilled is not better suited to our needs, as well as being more within our capacity to build and use."[103] With the Basic Industry Strategy, TANU now explicitly aimed to restructure industries around "increasing domestic linkages in order to achieve a greater degree of economic self-sufficiency." The plan sought to develop Tanzanian industries for producing iron and steel, industrial chemicals, farm implements, construction materials, and electricity along with a set of practical and modest consumer goods.

SIDO's Indo-Tanzanian Programme demonstrates how one of the key aspects of rethinking the technology of industrialization was to consider where the technology originated.[104] India's government was by the 1970s moving to domesticate manufacturing and production themselves, hoping to also make the country into the largest exporter of technology to the Third World. By 1975 Tanzania became India's most important market for capital goods such as machine parts, farm implements, power looms, and food processing equipment, all aimed at growing the small-scale industry sector.[105] Beyond the state's philosophical commitment to South-South economic cooperation, what made this new trade relationship enticing was that Indian firms could prove that their machines and tools worked well in tropical conditions. They were also appealing because the "rate of obsolescence is much slower in India." After a decade or two, spare parts for manufacturing machines would still be readily available. In other words, the stewardship of these machines would be far more possible than that of those imported from the West that became more rapidly obsolete. As the economist Sanjaya Lall has suggested, these growing relationships between India and other Third World countries actually cultivated a comparative advantage for India's outdated technology: "In some cases, . . . its selling point is its outdated, simplified, small-scale or adapted nature."[106] There was value not only in reusing materials but also in the machines themselves.

India was also an appealing origin for technology transfer as a leftist critique gained traction in Tanzania that, as the *Daily News* put it, "technology is not neutral."[107] Inseparable from the adoption of Western industrial technology, argued these critics, was the emerging critique that their modes of operation were unsustainable and required ultimately undesirable patterns of Western consumption inappropriate for socialist Tanzania. In response to this notion, the playwright Mukotani Rugyendo penned a blatantly allegorical play in 1977 telling the story of two men fighting over a woman named Maendeleo (Development). One of these men comes from a tribe that focuses on "individuality, decadence and the dependences on expatriates for their development soon after the links with the colonizing powers had been loosened" while the other comes from a tribe that "go[es] for real 'Maendeleo.'" This tribe tries instead to "breed their own experts, they establish people's institutions and cut off links with foreign experts. Foreign experts are attached to their brother (the colonizing power) robbers and plunderers."[108]

Reading Tanzania's industrialization plans and programs such as the Indo-Tanzanian Programme paints a picture of a nation even as early as the late 1960s trying to carve out a path to development that sought to avoid inappropriately expensive and hard-to-manage technologies. SIDO particularly captured the hopes that technology could be more decentered across Tanzania, shaped around labor reserves rather than unavailable capital reserves, and run by Tanzanians rather than foreign experts. In practice, however, time and again it seems that parastatal managers and government officials chose the opposite even when given a fairly straightforward option between the two. The result was reinforcing a centralized, capital-intensive manufacturing sector.

This happened for several reasons, including competition for production among parastatal managers who perhaps imagined that bigger was always better. It also reflected a general bureaucratic preoccupation with being as "modern" as possible. But it was also due to the nature of foreign aid. Bilateral aid packages often required the receiving country to purchase the technology for a project from the donor country, offering zero- or low-interest loans.[109] It was simply easier to "attract foreign finance for new projects—for this involves the winning of lucrative orders by firms in the donor countries."[110] These aid packages were by the late 1970s virtually the only way that Tanzania as well as many other developing countries could pursue its industrialization strategy any further, despite the fact that Nyerere openly bemoaned the inefficiencies and waste that accumulated at these sites of production.[111] Essentially, many decisions about technology in Tanzania came down to accepting what was on offer, which was routinely more costly, centralized, and "modern" than the state industrial policy advocated.

The resulting sugar refining plants or textile factories acquired through foreign aid tended to be what were called "turnkey" operations, with everything brand-new and fully provided upon delivery. In an apt extension of the "turnkey" language, these factories frequently became what was crucially called "technological lock-ins": sites of "packaged" technology that required buyers to not just purchase the initial technology but follow strict rules of where they could purchase spare and replacement parts.[112] As one critic pointed out, these sorts of black box technologies were problematic: "How can indigenous technology develop under such terms and conditions of imported technology?" Private local operations, on the other hand—usually owned by Tanzania's Asian community—far more readily turned to India for machinery, often buying secondhand.[113] However, even the technology choices under the Indo-Tanzanian Programme tended to be capital-intensive.[114]

Ultimately, many of the projects under the Indo-Tanzanian Programme that developed as alternatives to the Western development model also did not have great track records of success and much of the time the reason was, ironically, waste. Machinery sent from India sometimes arrived incomplete and without containerization, which could lead to damaged goods sometimes "dumped in the open" where they could be stolen or left to rust.[115] Spare parts could also take an agonizingly long time to arrive. The manager of the perennially delayed Kagera sugar plant in northwestern Tanzania stated, "Developing countries cannot help each other; they are too poor to provide the necessary service. So it is better to go for quality from the developed countries."[116] For many bureaucrats, though, the shared struggle against poverty, climate, and lack of expertise made India the far better choice for technology transfer. They saw Indian industries as more naturally attuned to the need to steward objects that were cheaper and easier to maintain and, when necessary, reinvent.

By the 1980s, when shortages were the most acute, promoting the progress of SIDO and the need to foster local raw materials had become a common news story and political message. Tanzanian engineers, Nyerere noted, needed to be creative and practical and open to "appropriate technology" as something that was not "ugly, temporary, and hard to use." They would need to develop locally sustainable maintenance systems and "importation should be our last resort."[117] Many factories in Dar endured years without any access to foreign exchange from the state with which to purchase crucial raw materials from abroad. At least on paper, the Tanzanian state had been anxious to rethink technology for decades before the economic crisis of the 1980s. But aid, competition among parastatals, and perhaps also the "class interests" of TANU bureaucrats had instead reinforced capital-intensive industries.[118] Now, with no foreign exchange, factories had to reconsider their machines and materials.

Facing extreme shortages, factories were sometimes forced to shut down for weeks or months at a time while their managers sought out practical and inventive solutions. The Kibo paper factory appealed to individuals and business to not burn wastepaper or throw it out but to call the factory to pick it up, since it could be used to make packaging materials, otherwise in dire shortage.[119] Local plastic firms could no longer rely on imported raw materials and were now ordered by the Ministry of Industries to make their crates from local wood resources instead of plastic.[120] This order emerged as a result of the ministry organizing a committee of experts to find methods for "reducing dependence on imported raw materials, capital machinery, accessories and spare parts."[121] Two years later, Simba Plastics Company, the largest plastics manufacturer in East and Central Africa, was still surviving after three years with no foreign exchange. Instead, they had begun collecting used plastic from industries and individuals and in 1984 they produced 572 metric tons of products from plastic waste and the dwindling remaining stock. The company's director pointed out that the factory was not, however, falling into disrepair when they did not have anything to manufacture: the machines were "oiled regularly and operated only when enough plastic wastes had been collected for use as raw materials."[122]

Dar's aluminum factory, Aluminum Africa (ALAF), also inaugurated a widespread salvage campaign to boost its stock. Faced with otherwise shutting down, the plant worked to locate twenty-three thousand metric tons of scrap metal yearly to produce its products, which were crucial for the construction industry. Dar's city council also gave the company permission to collect any abandoned motor vehicles carts and wagons across the city for conversion into scrap metal. When ALAF had conducted the same campaign in 1982, they had netted 150,000 metric tons of scrap metal around the city.[123] Dar's textile industry was also encouraged to explore new local resources for dyeing textiles, such as the bark of mangrove trees in Rufiji.[124] In the meantime, SIDO's director urged small industries to recycle "industrial and agricultural waste."[125] Both large and small, factories and artisans continued to muddy distinctions between formal and informal, new and used, stepping in and out of both realms as it suited them, taking objects in and out of the waste stream in the process.

Sabotage

Despite efforts to forge new channels for the production of food and essential consumer goods across the city, a prolonged sense of generalized want pervaded the city in the early 1980s. While there may have been multiple outlets for procuring goods and creative strategies for confronting shortages, these tactics took up an immense amount of time and delivered only enough to get by. Those still in formal jobs began to deliberately shorten their days "on the new pretext that they had to look for some essential commodity."[126] For frustrated and exhausted urbanites, three culprits emerged as culpable for this persistent state of shortages. The first was the president himself and his party loyalists: Nyerere was pressured but unwilling to negotiate with the IMF over market liberalization. As a result of his intransigence, foreign governments curtailed their aid to Tanzania at the behest of the fund. The state administered its own austerity policies aimed at urbanites to try and recover from debt, but this only further strained urban livelihoods and cut maintenance of infrastructures. Many Tanzanians were frustrated that Nyerere remained unwilling to compromise and felt that as a result the economy and the city were being driven into the ground.[127] The second culprit, one pointed out regularly by the president, was the IMF, representing an unchanged world order. According to Nyerere, if he complied with their demands to open Tanzania's markets and devalue the Tanzanian shilling, ujamaa would not survive. He believed that it was the obligation of any nation to resist surrendering "its right to restrict imports by measures designed to ensure that we import quinine rather than cosmetics, or buses rather than cars for the elite."[128] Indeed, Tanzanians might have agreed with this assessment politically while suffering from it economically. Many were frustrated both by the president and by the reigning economic order: these two perspectives were not necessarily contradictory. The third culprit, though, was one that citizens and the party could coalesce around: the growing numbers of Tanzanians considered "economic saboteurs."

By the 1980s the term "economic sabotage" was used to criminalize a broad range of activities such as illegally selling goods, smuggling goods across borders, and hoarding goods with the intent to profit from driving up their prices through scarcity.[129] With the passage of the Economic Sabotage Act in 1983, it became illegal to keep hidden any inventory of scarce products, and those caught faced possible imprisonment.[130] By this point, a cadre of businessmen and bureaucrats were making a decent living off the margins they could make by bringing goods across borders (arbitrage) or hoarding goods in order to cause scarcity and then profit off of higher prices.

TANU's aggressive campaign in the wake of the Sabotage Act failed to acknowledge different magnitudes or methods of sabotage, instead choosing to cast a wide net to catch as many purveyors of illegal goods as possible. In April, the first month of its enforcement, over four thousand people were arrested and by July 816 private shops had been closed down on suspicions of hoarding.[131] Prime minister Edward Sokoine headed the campaign against sabotage, and in the early days of announcing its implementation his speech-

5.3. "Annoyed city residents come to buy hoarded bed sheets from a shop that claimed to have imported them in 1975." *Daily News*, April 15, 1983.

es incriminating saboteurs drew large crowds as people anticipated the relief that would come with the capture of racketeers and the resulting new abundance of goods.[132] TANU and the press described the campaign as a triumph over chaos, lawlessness, and a polluted, diluted state: a victory over encroaching "capitalist elements in the country" that would restore the hope of socialism.[133] It was also, at least rhetorically, an exercise in disciplining the party itself. TANU members were in many cases the biggest saboteurs of all, using their political connections to stockpile goods or engage in other forms of corruption.[134] The persistent state of bare market shelves had allowed a politically connected and mobile class of Tanzanians to get quite rich, while the rest were unsure if they would ever again find basic goods again in regular stores. "It had come to a situation," one woman anonymously told a newspaper reporter, "where I never imagined . . . buying toothpaste in any open shop again. We had felt that there was no more effective steps which could arrest the situation."[135]

TANU and local newspapers made public theater out of the discovery of caches of goods. As the campaign got under way in April, picture after picture of recovered goods were splashed across the front pages of papers along with mug shots of captured "saboteurs."[136] In an arrest of thirty-five businessmen in Dodoma and Dar es Salaam, government officials described the hoards of goods to the media as looking "like Aladdin's Caves."[137] At a rally of fifty thousand people Sokoine also presented one hundred thousand Tanzanian shillings to a police officer who had refused to accept the same amount as a bribe from a "businessman" during the expulsion of racketeers.[138] The daily accounting of goods recovered offered a tantalizing and infuriating list of

5.4. A cartoon published in the *Daily News* April 12, 1983, of a hoarder trying to dispose of his money and goods while a policeman looks through his window.

what most could not obtain and had not been able to for years. When a Moshi businessman, Raphael Rango, was arrested, the *Daily News* detailed his stash of cash totaling 174,000 shillings along "with an assortment of goods—rice, sugar, cooking oil, wheat flour, hoes, iron sheets, 351 cartons of bulbs, beans, 615 bags of cement, 900 bags of maize, boxes of matches and paraffin—believed to have been acquired unlawfully from a nearby country."[139] Through this political theater, the sabotage campaign was not just about the recovery of goods but the recovery of trust in the state as the best providers of necessities by showing illegal sellers to simply be agents of suffering and exploitation. Paired just a month later with the beginning of Nguvu Kazi, ridding the state of saboteurs and the city of "unemployed" were essentially campaigns to "clean up" the nation and restore faith in the socialist project. TANU realized that this was impossible without reclaiming control over the channels of production and consumption. It is in this context that we also have to consider the devastating blow to TANU when Sokoine died some months later, in 1984. The heir apparent to Nyerere and the face of the campaign to restore order to the state, Sokoine was killed in a car accident on his way from Dodoma to Dar es Salaam. Rumors swirled around his death that it had been plotted by saboteurs who feared the return to law and order.[140]

But as paralyzing as hoarders and racketeers could be to the movement of goods on the market, they also sometimes served as essential conduits of distribution for the wanting city. Prior to the campaign, illegal marketeers had even frequently bought directly from factories, importers, and wholesalers, functioning as one of the major pathways for getting goods to consumers. While the campaign promised and briefly led to full shelves again at local shops, by sealing up the leakages and alternative channels of the market, it also scared many people into hiding, closing up shop. Ultimately, TANU's pursuit of walanguzi failed to anticipate its accompanying consequences. Cracking down across the board on illegal but critical purveyors led to, as the *Daily News* put it, "the narrowing, if not outright blockage, of outlets and channels."[141] Some goods completely disappeared and for others people were forced to pay even higher prices at "more inaccessible markets."[142]

Another dramatic side effect of the campaign was the need for quick distribution and disposal. Authorities were faced with sorting through perishable goods as quickly as they could and selling them to prisons, schools, village shops, and the army before they spoiled. The city council took to destroying unearthed bags of wheat found in Wazo Hill and containers of expired milk while regional authorities turned food that "was no longer fit for human consumption into animal feed."[143] With the specter of being caught, many saboteurs also decided to simply dispose of their caches however they could—rumors circulated of goods being spilled into the streets of Dar es Salaam as well as thousands of liters of diesel being dumped into the bush near Dodoma. As T. L. Maliyamkono and Mboya S. D. Bagachwa write, "During the first few days of the crackdown one was likely to be offered bundles of banknotes by guilty people anxious to be rid of them. Those who were too scared to carry their money to the banks threw bags of notes on the road, but the government hastened to create an anonymous money dumping place. Ivory waiting for shipment was buried. The rich became grave-diggers of the 'tombs of wealth'—televisions and videos which had probably been smuggled into the country."[144] Nyerere exhorted people to not throw things away but to surrender them for amnesty.[145] In attacking illegal means of distribution, the government unintentionally caused waste to abound at a time of acute shortage.

While these problems of disposal proliferated, TANU nevertheless conjured their campaign in metaphors of dirt and cleanliness. Sokoine had announced to audiences that getting rid of saboteurs would help "build a socialist and self-reliant society of morally clean people."[146] Nyerere claimed this new law would also be a "thorough re-examination" of party members

5.5. Cartoon of exiled urbanites joining agricultural farms outside of the city. *Daily News*, 1983.

and the government "with a view to cleansing themselves by identifying all those who are corrupt and reporting them to the appropriate authorities for appropriate disciplinary action."[147] Cleansing the party and Tanzanian society of corruption also required a return to the land. To reclaim ujamaa, those caught would be given hoes and made to till the soil: "That way, they will be liberated and cleansed and will be made to understand good living was through work and sweat."[148] Sabotage, accumulation, and waste were part of an urban landscape that was corrupting as well as parasitic.

The war against saboteurs, when considered along with the era's frequent campaigns to clean up the city and expel the unemployed, was an effort to make the city and the economy legible again from above while also imposing notions of what kinds of productivity were moral. Nyerere frequently reminded Tanzanians that saboteurs threatened to derail the project of socialism. In its most literal definition, *sabotage* means to render something inoperable; to impair or cripple. And in many instances Tanzania's hoarders and racketeers were indeed rendering the economy inoperable. But there were also many other unintentional agents of sabotage that shaped Dar's economy and productivity. For example, in 1984 the mud accumulating in the Lower Ruvu Valley's water treatment plant had come to take up six out of seven meters in the holding tanks, making it impossible to pump the necessary amount of water for servicing a good portion of the city. As an unwitting agent of sabotage, this mud shut down the paper plant, cigarette plant, aluminum plant, and milk production facilities until it could be cleaned out (only to fill up once again).[149] Not unlike rain stymying the arrival of buses, industrial saboteurs took many nonhuman forms, as the city's clogged channels and new flows took on a life of their own. When such problems of environmental sabotage occurred, it was frequently the illegal sellers of the city who were best equipped to supply crucial alternative channels of distribution. It was both real and metaphorical dirt and waste that continually threatened the authority of the state.[150]

This chapter has sought to bring several narratives and events together, perhaps at times too much. But the scope of this and the previous chapter is quite purposefully large. It represents the fact that for postcolonial nations, economic and infrastructural problems rarely remained isolated or contained. One bad growing season could have profoundly reverberating effects. In moving from food to waste to the dilemmas of technology in industrialization and finally to sabotage, I seek to bring the material repercussions and transformations of Dar's landscape together with the political dilemmas of how to produce goods and control waste in a nation shaped by scant infrastructures and dwindling foreign exchange. If navigating profound shortages is likely how most Tanzanians would remember this period, waste is nevertheless the hinge at the center of these stories. Dirt, garbage, and toxic industrial flows marked the city and yet also signaled new productive urban landscapes. Seeking ways to limit wasted time, money, and precious resources prompted new approaches to industrialization; Dar's factories had little practical choice but to creatively reassess their own resources and practices. And finally, as a rhetorical device to police certain populations, the state took up narratives of dirt and cleanliness to prosecute loiterers and saboteurs and reassert their moral authority over consumption and production.

Yet for all the finger pointing by the state, this chapter has also shown how willing regional authorities were to actively blur distinctions between wasting and wanting, urban and rural, formal and informal. Reanimating the city's resources required reimagining the relationship between Dar, its citizens, and the region's environment. The new descaled sites of production that emerged in the city's interstitial spaces were ideological and practical.[151] In the eyes of the state, urbanites should do what they could to sever their parasitic relationship with rural Tanzania as well as foreign imports. What counted as wasted spaces in the city were unproductive spaces, not necessarily the same areas where trash and mud accumulated or toxicity accrued.

The literature on African urban economies and spaces in the 1980s and since has focused on the massive expansion of an alternative "informal" economy that emerges, frequently at odds with the state. Its rise is marked by a corresponding decline in formal wage labor. Yet by examining state efforts to provoke local production of both food and manufactured goods, this chapter shows, as other scholars have also pointed out, that what has come to be labeled the "informal economy" did not evolve in the shadow of the formal economy but rather frequently hand in hand with it, sometimes exploiting the same methods of distribution and spaces of production. In many instances, new modes of production in the city—whether in industrial workshops, under trees, or in backyard gardens—were explicitly encouraged by the government with the intention that they would become formalized or simply disappear after serving their purpose as temporary triage.

In noting how present the state was in these enterprises, this chapter and the next broadly suggest a need to consider a more historically accurate distinction of the emerging dual economies of the city. In the case of Dar, it seems that they were distinguished not by formality but by dependencies on oil and foreign exchange. Indeed, this distinction helps illuminate the nature and limits of Nyerere's antiurbanism. The state defined acceptable productivity in the city in two ways: either by participating in an economic activity that lessened the need for foreign exchange (which included planting food and crafting new objects that lessened the need for imports) or by working in an industry that created foreign exchange through producing goods for export. In this way, supposed loiterers who mostly made a living through exploiting small margins in the marketplace of the city and seemingly produced nothing were, as Nyerere opined in 1983, equally as destructive as economic saboteurs.[152] This is the same logic that allowed the city council to frame compulsory farming projects in the city's polluted valleys as productive while demolishing the stands of unlicensed vendors in the name of public health.[153]

Urban communities, however, had their own metrics of productivity, sovereignty, and well-being. In the midst of the 1983 Nguvu Kazi campaign to expel unemployed urbanites to regional farms, a middle-aged vendor selling groundnuts explained to a newspaper reporter that he bought groundnuts at wholesale prices and after roasting them would resell them for a modest profit. When the interviewer asked him why he didn't farm instead, he answered, "If I grow crops what would I be looking for: food and money—alright? What is the difference if I get the same through groundnut selling"?[154] A fish seller also told the interviewer, "I don't see how you will convince me that a certain job is better than what I am doing now. Even if you give me employment in your company, I won't accept it. . . . I just can't survive on a monthly salary." While there was no reason for urbanites to necessarily subscribe to the state's conception of productivity as described above, they *were* compelled by circumstance to seek out economic activities that also sidestepped foreign exchange. In this way, whether tacitly sanctioned by the state or not, Dar's emerging economy was very adept at appropriating the leftovers of the "modern economy" through refashioning its passive byproducts and scraps but also through the "driver who sells cheaply some of the petrol from the official car, or provides clandestine transport; a junior messenger who supplies old paper for wrapping to small shopkeepers, the maintenance technician who takes his work tools to use during non-working hours."[155] The line between hoarding goods and the stewardship of objects could be overt, but it could also be quite hazy, with the former considered sabotage and the latter a good socialist ethos.

Chapter 6

FUELING CRISIS

Charcoal, Oil, and Economic Sovereignty

As the shifting contours of wasting and wanting redrew the city's outer edges, the flurry of land clearing for suburban housing, regional ujamaa villages, and agriculture also created an increasingly lucrative byproduct: charcoal. In an era of pervasive urban shortages, perhaps no commodity better captured the city's growing reliance on its hinterland. But if the production of charcoal came to prop up a vital local economy, it was also increasingly entangled in global anxieties about energy, environmental crisis, and reappraising the terms of international development.

Charcoal is an incredibly adaptive form of energy that has long been central to Dar's households as a primary cooking fuel. Yet as the price of "modern fuels" (petroleum and kerosene) skyrocketed first in the 1970s and then again in the early 1980s, the role of charcoal expanded to also become a primary fuel of the new urban economy. In Dar's sylvan periphery, urbanites found ways to deflect the vagaries of high energy prices through selling charcoal to supplement their wages and designing small businesses around its availability. Charcoal's increasing use reflected the scramble to find alternatives to "modern fuels" and also signaled Tanzania's continued rapid urban growth. Whereas only a small percentage of rural Tanzanians used charcoal (gathering firewood instead), 85 percent of urban Tanzanians relied on it in some form, even if in combination with other fuels.[1] Compared to firewood, charcoal is relatively smokeless and burns at a higher temperature, making it ideal for tight urban spaces that sometimes lack ventilation.[2] It was, and today remains, a quintessentially urban fuel.

To a growing number of international environmental organizations and researchers in the 1970s and 1980s, charcoal represented something quite more sinister than merely an alternative to "modern fuels." Its expanding use and the corresponding incursion into Tanzania's forests to produce it became part of a new global threat. On the eve of "sustainability" emerging as the new keyword in international development, wood fuel (charcoal and firewood together make up this category) garnered unprecedented attention as part of an alarming environmental crisis. Sparked by neo-Malthusian concerns of surging populations, environmentalists and ecologists in the 1970s argued that the Third World's overwhelming reliance on biomass for energy had created "The Other Energy Crisis" alongside the oil crisis.[3] This crisis, observers worried, would culminate in global deforestation in a matter of decades. While for much of the postcolonial period economists and social scientists were concerned that the continued expansion of African cities would erode the reservoirs of labor needed for viable rural economies, the continent's cities now also posed an existential threat to rural and global environments. Continuing a long legacy of pitting the city against the country in African development narratives, charcoal use placed fast-growing cities like Dar at the "heart of deforestation in Africa."[4]

This new era of environmental crisis looked different from the perspective of the Global South. For many Third World leaders, exploiting and exporting natural resources was key to increasing their capacity to reach economic sovereignty. For those leaders, including Nyerere, the fuelwood crisis was not parallel to the politics of oil but rather inextricably embedded within it: all energy use reflected unfair terms of trade and austere conditions of international development. Wood was not a fuel choice divorced from the price of oil or the fluctuating prices of Tanzania's commodities on the global marketplace. While its use threatened Tanzanian environments, it was also renewable and eased the necessity of importing fossil fuels. By 1982 Tanzania was spending 60 percent of all its export earnings on purchasing petroleum to the detriment of buying spare parts, new machinery, or fertilizer—the prices of many of these crucial imports were themselves tied up with the petroleum economy. Rising oil prices also spiked the price of other "modern fuels" such as kerosene and diesel, which powered a good portion of Dar's electrical grid. And yet the petroleum imports that took up more than half of Tanzania's foreign exchange accounted for only 7 percent of the nation's total energy use.[5] By the 1980s biomass made up 98 percent of the energy used by Tanzanians in their daily lives and the percentage was only increasing. Wood fuel was not its own crisis so much as a symptom of the ongoing unequal terms of trade. In this regard, Tanzanians did not look very different from most members of the Third World.

What makes charcoal and its energy-intensive production such an important part of Dar's environmental history, then, is both its real and discursive circulation and use. This chapter weaves together local responses to the reverberating effects of a globalized crisis (the oil crisis) with international responses to local resource use (the fuelwood crisis). Together, these interlocking discourses of crisis shaped the state's viable futures in the late socialist period into diverging development futures: one of "sustainable development" shaped by international interventions and lending conditionalities, and the other what I call a "grounded" development shaped by technological and resource solidarities across the Third World.

In this urgent context, charcoal meant many different things to many different people. For international conservationists, charcoal signaled a reason to intervene in Tanzanian development policy, this time with a new language of sustainability and biodiversity. For the Tanzanian state, charcoal became a potential path to resource sovereignty, both modeled after the political example of the oil embargo and a practical reaction to the embargo's devastating economic aftermath for non-petroleum-producing nations of the Global South. The charcoal economy became key to the state's search for an alternative to the increasingly straitjacketed terms of aid and externally imposed "sustainable development." For Dar's communities of charcoal producers and users, the fuel meant something far more elemental. Charcoal offered a lucrative way to participate in the urban economy, particularly as agricultural commodity prices plummeted. And thus while the international environmental community and the state wanted to regulate it, each for their own ends, local communities found that charcoal could function as an anarchic fuel source. Charcoal became key to navigating both state and internationally imposed austerity. Suggesting that charcoal served as an anarchic fuel source is not, as I will show, to suggest that charcoal production was chaotic or undisciplined but rather that it was ultimately impossible for the state to adequately regulate it, precisely because of its materiality and modes of production.[6]

The international consensus of the Third World's fuelwood crisis emerged through the extensive production of ecological and social scientific case studies. This gray literature makes up the bulk of my sources in this chapter as they trace the contours of the crisis narrative, chronicle interventions, and capture the itineraries of Dar's charcoal market. These studies also became a crucial way of knowing and governing people and forests. There are actual roots at the center of this chapter—uprooted trees and their transformation into charcoal—but I am also interested in how metaphors of rootedness were used to insist on tree planting as a global salve not just for environmental problems but for uplifting local communities and particularly women.

Where previous chapters have focused on the ways in which Dar's built environment was shaped by tensions between the city, its residents, and the politics of ujamaa (both national and international), this chapter moves between Dar's woodlands and forests and a coalescing international discourse on deforestation. In this literature, cities initially play a peripheral role within debates over fuelwood, which is first identified as a crisis of rural communities in arid regions. Yet I argue that over the course of the decade of the "fuelwood crisis" and certainly now, the growth of African cities and their continued reliance on charcoal emerge as the most imposing threat to forests and the continent.

Forests and Charcoal in Coastal Tanzania

Coastal Tanzania's diverse woodlands and forests have historically met a plethora of local needs while also remaining central to the larger Swahili coast's precolonial economy. Sixty-six forests dot the Tanzanian coast from Mozambique to Kenya, including eighty-thousand hectares of coastal mangroves. And while quantifying forests is a particularly fraught practice I will return to later, by some estimates nearly half of Tanzania is covered in either "closed forests" or the more scattered and open canopy of savannah or *miombo* woodlands.[7] Before the colonial era, coastal forests were the heart of the copal trade between the mainland and Zanzibar, enriching local Zaramo chiefs who controlled access to forests as "wielders of the ax."[8] In addition to a source of copal, and the occasional ivory, these forests also served as a refuge from warfare, a place of spirits, a space for farming, and a source of income by demanding tolls from caravans traveling between Lake Nyanza to the coast.[9] Forests also supplied trees for the Swahili coast's venerable woodcarving tradition as well as small fishing vessels known as dhows that have traveled the Indian Ocean for centuries.

With the arrival of colonial rule, who was allowed to use forests and how began to rapidly change. Because the history of forestry in Tanganyika has been covered quite well by a number of scholars, I will not delve deeply into it except to identify two themes.[10] First, a self-evident but important one: over time, forests have been defined differently depending on the imperial or state interests in who should or should not use them and what purposes they serve. What counts as a forest is an acutely contingent question, bound to place, time, and prevailing power relations. Local communities unsurprisingly defined them and used them differently than what the colonial state tried to legally enforce. As a result, colonial forest history is animated by conflicts between state and local aims and definitions. Secondly, in moments when European foresters sought a greater role in gatekeeping forests, they were also invested in exploiting them efficiently and maximally. The former meant keeping out peasants who farmed, foraged, cut trees, and made charcoal, but the latter required the presence of a local labor force.

This tension between exclusion and the necessity of labor came to a head in the 1950s, as global lumber prices rose. The number of hectares designated as forest reserves in Tanganyika grew dramatically to enclose fourteen times the land of prewar amounts.[11] In many rural areas, the British also began anxiously monitoring how peasants used Tanganyika's ungazetted forests. With the Forest Ordinance of 1957, it became illegal to fell trees without a commercial license, while the state could designate forest reserves on any otherwise "unclaimed" land. Accordingly, if "any person who without a license or other lawful authority cuts, fells, damages or removes any tree on unreserved land for the purpose of sale, barter, or for use in any trade, industry or commercial undertaking shall be guilty of an offense under the Ordinance."[12] In response to the increasing power of colonial foresters, TANU leaders began speaking out against the tyranny of conservation efforts as a key harm of colonialism. Independence movements took up sometimes violent anticolonial tactics in response to authoritarian measures to curb erosion and limit grazing.[13] Thus, there is an important history for many Tanzanians of being kicked out of forest reserves and a palpable resentment toward state authorities who try to police their borders and control how they are used.

In the areas around Dar, however, as forest reserves were expanding in the 1950s, the colonial state sought out and accommodated forest squatters.[14] Not only were they necessary as a labor force but Dar's forests offered a potential antidote for rapid urban migration. Similar to concurrent efforts to create villages as agricultural satellites to keep Africans out of swiftly growing urban neighborhoods, forest reserves offered a solution by settling squatters in reserves where they could cultivate their own agriculture, harvest trees, and plant new hardwood and softwood species for colonial timber operations.[15]

This relationship between authorities, forests, and locals outside of Dar meant that communities were allowed to engage in charcoal production as a crucial local fuel. Furthermore, charcoal production was categorized as a subsistence activity that used wood not earmarked for export and did not significantly interfere or compete with other forest products. In fact, by the end of colonial rule, colonial administrators encouraged Africans in some places "deeper into the forests" to collect copal and beeswax, tap rubber, build canoes, and strip mangrove bark so that they could pay their taxes.[16]

The colonial state was also likely permissive of forays into the forest for charcoal because it was a way that communities provided for themselves. Peri-urban charcoal production allowed families in the city to meet their own fuel needs. The use of forests around Dar—particularly for fuel—lessened the burden of erecting urban infrastructure and expanding the reach and use of more expensive fuel sources. In a climate where heating needs were otherwise negligible, cooking fuel was the primary energy use of African families and it could be easily met in Dar's hinterland without state intervention. In this way, charcoal and firewood functioned as tools of underdevelopment. Forests "provided a subsidy in nature" that allowed the state to then avoid investing in the energy infrastructures that would be necessary to replace its use.[17]

One important caveat to the relative freedom to conduct subsistence activities in the region's forests was the colonial state's postwar campaign against miombo woodlands. The British came to see the region's sparser woodlands, previously not considered forests, as containing newly valuable trees for exploitation as well as the battleground in the fight against the tse-

tse fly. Without proper maintenance and intervention, the state saw these woodlands as an unwelcome bulwark against "civilization" and progress. Furthermore, the local practice of using fire to open new farmlands for shifting cultivation "elicited a colonial hysteria that peasants and pastoralists . . . threatened to dessicate the continent."[18]

These changing policies capture the larger history of ambivalence about how communities used forests, an ambivalence that continued in the postcolonial period. Importantly, it demonstrates how the colonial penchant for permissiveness and access for subsistence practices could shift quickly to a deeply paternalistic and martial policing of forest boundaries. As a result, Africans were, on the one hand, obliged to always keep "one foot in the subsistence economy" by producing charcoal. But on the other hand they were expected to increasingly leave forests alone for state accumulation.[19] Peasants could quickly become invaders in the eyes of the state, incapable of conserving trees and stewarding forests.

In the early years of independence, the crucial relationship between the city, peri-urban communities, and forests only grew. Charcoal production was even sponsored by the World Bank, who saw forest reserves as a way to engage the "maximum amounts of labour from Dar es Salaam" and help mediate the problem of growing unemployment in the city.[20] By the late 1960s the state wanted to gain more control over charcoal making, not because it feared the effects but because it sought to modernize and increase production, selling it on international markets to generate foreign exchange.[21] State foresters predicted that if charcoal operations were both expanded and made more efficient, they could employ at least five thousand people.[22] Ironically, it was predominantly fuel-rich Middle Eastern countries that sought charcoal from East Africa, but other countries, including Germany, also expressed interest if the charcoal was of high enough quality.[23]

Tanzania Charcoal Manufacturers Limited (Tancoal) sought plans to establish a plant that could produce twelve to fifteen thousand metric tons of charcoal annually with further plans to produce charcoal byproducts such as settled tar, acetic acid, and methanol.[24] This new capacity, Tancoal suggested in its proposal, would be of "great importance" because it "offers opportunities for employment, utilization of local resources, foreign exchange earnings and general development and diversification of the economy." In essence, many in the state and private industry saw charcoal as a path to development. In 1971 Tanzanian producers sold 188 tons of charcoal internationally with the hopes of undercutting Kenyan prices and expanding the market to at least three hundred thousand tons annually in the following years.[25] The Forest Division also claimed to have taken over charcoal production in many areas, supervising burns and buying it from the locals to sell to town residents. They hoped that by organizing "big scale charcoal production near all bigger

6.1. Clearing for an ujamaa village in the coastal region of Tanzania, 1972.
Photo courtesy of Tanzania Information Services.

towns" they would not only "satisfy local demand for this kind of fuel" but also render it unnecessary to "import fuels from abroad."[26] Observers noted, though, that it was unlikely that the state exerted any real control over charcoal production and its scope.[27] The United Nations Industrial Development Organization also sponsored a meeting between Uganda, Tanzania, and Kenya to discuss the prospect of an East African charcoal industry.[28] Charcoal had shifted from a subsistence activity to something that the state imagined could generate foreign exchange. And yet it is not clear what happened to these efforts, because the industry did not emerge as imagined.

In the 1970s forest boundaries attenuated and incursions into them increased. This was due partly to the state's efforts to expand commercial forestry and promote wood as a "modern" building material, but it was also due to the massive increase in ujamaa villages that reshaped the nations forests dramatically at this time. The collapse of food distribution channels and the long drought also prompted regional authorities to roll back their efforts at keeping communities out of forests.[29] Meanwhile, skyrocketing oil prices only further magnified the importance of both charcoal and producing food locally.

The permissiveness of the state and the urgency of generalized shortages spurred deeper entanglements between Dar and its hinterland. Charcoal production began to transform the urban economy and the peri-urban environment. While biomass harvested from local forests clearly could not replace petroleum as a fuel source in a car or most factories, as petroleum became harder to procure a charcoal economy expanded to take its place. And as charcoal gained value and settlement in the hinterlands became more intense, peri-urban villagers who still relied on firewood to cook their evening meal also found their fuel landscapes altered.

Making Charcoal

Across the latticework of forests and woodlands that dot Dar's periphery, peri-urban farmers frequently produced charcoal in their off-season. Local charcoal producers preferred the trees of miombo woodlands, which make up a vast part of the southern subtropic zone of the continent and abut the edges of the city. Less dense than the region's forests, these woodlands were easier to work in and frequently remained outside of the purview of state control after independence. Within these woodlands, charcoal producers also had a hierarchy of trees. What in Kiswahili is known as miombo, scientists refer to as the *Brachystegia* genus of trees, and the region is home to a broad range of species within this genus. Some of these trees are denser than others, which is a good trait for charcoal. Some *Brachystegia* also have more armor against the intrusions of local animals and humans: they sprout nightmarish-looking thorns. For obvious reasons, these trees tended to be less-preferable fodder for charcoal makers. Charcoal producers also took location into account, hugging the contours of Dar's expanding arterial roads or followed along paths of peri-urban settlement where trees were cleared for cultivation or villages, allowing easier passage to roads and markets.

After choosing trees for a burn, charcoal makers would dig a hole that would serve as a kiln and place the cut wood inside. They would then light a fire and cover the kiln with blocks of earth and grass, creating an environment where slow carbonization could occur with a limited supply of air. Over the course of a few days, the covered wood burned under conditions of very high heat and low oxygen. If they had filled it with dense wood, charcoal makers knew it could take around five days to transform into charcoal, but poorer-quality wood required less time or it might incinerate.[30] In choosing the wood, monitoring the fire, and extracting the final product, a charcoal producer hoped to incur the least amount of energy loss. But even when done right, it could still take about one hundred kilograms of wood to produce between eight and twenty-three kilograms of charcoal in an earth kiln.[31]

Dar's charcoal makers used packed-earth kilns because they required virtually no input beyond local labor, materials, and time. In other words, these basic kilns had their own logic of efficiency even as they required an immense amount of wood for a small yield. In this transformative process, the resulting carbonized product unearthed from the ground returned to the surface as a shiny, black, and substantially lighter version of the trees cut and buried a few days before. The dramatic transformation was monetary as well as material. While the firewood that went into the ground had virtually no economic value, charcoal had a steady supply of urban consumers. What emerged was also an eminently mobile form of energy that was ready for use with very few other tools. This is also key to understanding its potential as an ungovernable fuel source. With little input, charcoal production could occur virtually anywhere in the forest and did not need to be scaled up to be lucrative for small-time producers.

The infrastructure of charcoal production also made it easy to do without the infrastructure of pipes or trucks otherwise necessary for delivering other energy sources. Once cooled, charcoal was packed into sisal sacks and brought into the city usually on trucks or by bicycle. A man on a single-speed bike heading over the rolling hills into the city could carry maybe a stack of bags three or four high strapped on the back of the cycle. The bike, itself requiring no "modern fuels" and only the occasional and often improvised spare part, was the arch infrastructure of small-scale charcoal distribution. Even when a particularly heavy load or a flat tire rendered a bike unridable over Dar's hilly topography, bikes could be walked and pushed into town.

Unlike a similar load of wood, however, a bag of charcoal was light enough to also carry as a head load.[32] If stuck without a bike, charcoal sellers could walk out to a main road and wait to hitch a ride or catch a passing bus or make the trip all the way into town. One study in 1983 found that an "elderly gentleman" in Temboni "brought a sack load of charcoal all the way from Kinyerezi, eleven km in the interior." After arriving in the city, he would sell the sack for forty-five shillings and "hoped to earn enough to purchase a few kilograms of sugar, maize flour and fish at Temboni market centre. There are many peasants of his description who occasionally resort to charcoal burning to satisfy similar needs. When a mango or cashew-nut tree no longer produces sufficient fruits, it is felled and is burnt into charcoal for cash to meet immediate needs."[33] Charcoal production was also quintessentially local and distributed, making it diametrically opposed to petroleum. Long, bumpy trips on roads pulverized charcoal, diminishing its value; the closer markets and forests were to each other, the better.[34]

Just as charcoal did not tend to be the driving force in cutting forests down, it was also a way to make money interspersed with many other opportunities and options. One informant told researchers in the 1980s that he made charcoal during the month of Ramadan because it took less energy than farming, and therefore it was easier to endure without food.[35] It was also lucrative enough to provide cash for the feast foods needed for the month. Charcoal production occurred on a variety of scales in as many locales. Sometimes neighbors or families would come together for a charcoal burn, cooperating on the tasks so that it could be done alongside other jobs or staggering multiple burns to create a steady supply.

As wages declined, urbanites turned outward to the forests to make up for stagnant salaries. For example, a retired schoolteacher started up a wholesale charcoal business to "put a little in the bank and just wait for time to pass" until he could "go and see *babu* [the old man] in heaven."[36] A charcoal dealer with a bicycle could on an ambitious day make three trips between a peri-urban charcoal distribution station and the city, earning "sh. 90/- a day which is untaxed. This is almost the net daily pay of a Tanzanian University Professor or Senior Industrial Manager in Tanzania."[37] Unlike wholesalers who kept their charcoal supplies in godowns, small-scale sellers who bought from them or produced smaller batches would sell charcoal in a few different-sized sacks—or, while paying a higher rate, customers could also buy charcoal by the handful to make a hot evening meal possible.

The charcoal industry in the periphery popped up alongside new neighborhoods. One such spot was the village of Kimara located along Morogoro Road in "undulating countryside, consisting of a series of hills and hillocks, interspersed with ephemeral streams and dry valleys." As the host of forces that animated urban life in the 1970s pushed Dar's residents outside of the city, the older woodland vegetation became interspersed with cashew, banana, and mango trees. By 1983 Kimara had become "rural Suburbia, a favourite residential area for the Dar es Salaam top bureaucrats, professionals and businessmen who are erecting modern houses on the villages salubrious hills and hillocks." They were methodically buying out the original inhabitants of the area and using their new land to farm and "cushion themselves against inflation."[38] These new bureaucrats of suburbia, along with more established residents, were using the forest's resources and the region's clay soil to make bricks and start a number of "market stalls, shops, eating houses, bars" along the road including two butcheries, six shops, four meat roasters, and launderette.[39] Kimara's new vibrancy in part reflected its proximity to a charcoal site, Bonyokwa, seven miles west from Kimara's bus station where by the early 1980s around seventy people were working full-time burning charcoal, with more taking up the work in off-seasons. Charcoal production was perfectly paired with the flurry of peri-urban settlement that felled trees in its path. While the expanding city gave rise to a larger charcoal market, land cleared for new homes, businesses, and ujamaa villages literally fueled the trade.

For local consumers as well as producers, perhaps charcoal's biggest advantage was that it could be distributed and used entirely outside the physical infrastructures that make up the delivery methods of "modern fuels," such as oil, kerosene, and electricity. It likewise required no foreign exchange, although charcoal distributors certainly took advantage of cars, buses, and trucks when they were available. Charcoal both relied on and created distribution networks that epitomize what AbdouMaliq Simone has called "people as infrastructure."[40] As explored in the previous chapter, the continual rupturing of service delivery by the late 1970s made many African cities run through "economic collaboration among residents" rather than "systems of highways, pipes, wires of cables."[41] Pipes, wires, and cables centralize and control distribution while also exposing delivery to the vulnerabilities of system disruptions and sabotage. But with charcoal, most urban residents lived a few minutes' walk from charcoal sellers and could buy as little or as much as they needed (those who bought smaller amounts did pay a premium) for the evening meal or to last through the next month.[42]

Gathering Firewood

For households in the periphery that were not in the charcoal trade, the fuel's new value and the expansion of the city could be more troubling than lucrative. As a woman living in the peri-urban ujamaa village of Mwanalugali recalled, it was in 1974 that "the shilling began melting like ice and our life patterns changed."[43] With falling wages, one way families on the city's fringes could save money was to stop using charcoal. This could dramatically reshape relationships between communities and their landscapes, particularly by changing the daily routines of women. When villagers first established Mwanalugali as part of Operation Pwani in 1973, the women of the village recalled using charcoal on occasion but finding firewood had been easy enough that it served as their primary source of fuel. In the early years of the village, it was lions more than wood scarcity that women feared when they ventured out to collect fuel. To protect themselves against wild animals they usually collected wood in groups, turning an onerous chore into a social outing. But soon women reported to researchers that they had to walk ever farther to collect enough wood to cook a meal as deforestation expanded. In the short term, charcoal production could offer women an easy way to collect wood because they could follow behind groups of "charcoal men" they could take up their remnants.[44] But the compounding pressures on forests in the region soon began to impose heavily on their lives. As one woman described the dilemma, "The amount of time that we are wasting collecting fuelwood, we could have accomplished a lot of chores at home. We usually spend about 4 to 6 hours in the forest. If I was at home, I could have accomplished a major task. For example, if I went to the field, I could have cultivated a reasonable portion."[45] While this estimate of time spent finding firewood seems almost absurdly high and likely was not a daily expenditure, recording the hours women now spent gathering fuelwood became a hallmark of fuelwood studies.

For these women, deciding whether to use firewood or charcoal for an evening meal was a calculation of time, money, and what food was available.

Beans took so long to cook that they fell by the wayside in times of fuel short-ages. On the other hand, vegetables and fish cooked quickly, which could also be an advantage when so much time was spent gathering wood.[46] During the rainy season, when charcoal was also more expensive and harder to produce, their alternative was wet wood, which rendered food smoky and it took for-ever to even boil water.[47] Another woman added, "There are times you would feel like switching from collecting fuelwood" but, she lamented, in doing so, "you will realize that you have disturbed the budget so much that you can hardly meet other needs such as *mboga* [vegetables] and maize flour/corn meal. Hence, we prefer to collect ourselves in order to save 400 Tsh for other needs."[48] For women, charcoal and firewood represented the frustrating daily process of deciding between the value of their time or their money.

The forests and woodlands around Dar assumed an auspicious dual role in the 1970s as a much-needed refuge against the surging prices of oil and kerosene but also, as we will now turn to, the site of growing concern in the emerging narrative of the global fuelwood crisis. Firewood and charcoal in the periphery also furtively marked out the shifting and flexible material boundary between city and country that carved its way through the land-scape. Unlike firewood use, which did not rise above the level of a family sub-sistence activity, charcoal provided a crucial supplementary income for those involved in its production and distribution. Frequently ignoring the multiple factors leading to increased reliance on wood, global attention was now be-coming affixed on how peasants turned trees into fuel. Experts as early as 1970 began to worry that fuelwood consumption had increased 35 percent in the past decade, leading to twice as many trees felled as fifty years prior. One study on desertification projected that if increasing incursions into the for-ests for fuelwood use were not halted, charcoal consumption would increase 1,120 percent by the end of the century.[49]

The Emergence of Global Environmental Problems

The narrative of the wood fuel crisis—both its discursive construction by the global scientific community and the research methods used to identify it—must be understood within the context of the emerging environmental movement in the early 1970s. This movement coalesced around new concep-tualizations of ecological problems on a planetary scale. At the helm of this movement was a growing number of voices that I refer to here as "Northern environmentalists" who began writing and reading a nascent literature on the earth's finite resources and the "limits of growth."[50] They created a new language to describe the emergence of a generalized environmental crisis and it quickly gained traction in public discourse. In just a few years, nearly every industrialized nation began implementing policies explicitly aimed at addressing issues such as industrial pollution and wildlife extinction; these problems were identified as persistent and pervasive threats not just to local communities but also to the planet as a whole.

The oil crisis also helped foster this emerging notion of a global environment. Beyond serving as an example of resource scarcity (or at least the orchestrated effects of scarcity), the oil industry in the wake of the embargo began actively promoting a particular notion of environmental conservation. While in the late 1960s the oil industry had been predicting and preparing for the end of oil reserves, this changed when the embargo effectively raised oil prices.[51] The specter of scarcity was no longer necessary to drive up prices. The oil industry instead began to worry that fears of finite supplies might lead to enough energy alternatives to render hydrocarbons obsolete. According to Timothy Mitchell, one method used by oil companies to ensure the continued relevance and value of oil was to "champion conservation and the protection of the environment" as a way to discourage the exploration of alternative fuels.[52] For environmentalists, who also began organizing around oil spills as well as the prospect of the end of oil, the finitude of natural resources helped frame and politicize "the environment" as "an alternative project to that of 'the economy.'"[53] The oil crisis was therefore key to the increased use of wood in much of the world and the new conception of finite global resources that made this new use of wood alarming. Forests needed to be saved in the name of the environment.

Turning their gaze outward to the Third World, the nascent environmental movement also nurtured a neo-Malthusian image of the Global South as a vast and ever-growing population of soon-to-be consumers and polluters.[54] Through the founding of new organizations and a series of international conferences, a vocal and influential group of scientists and activists became critics of existing models of international development. At the heart of their critique they urged governments and global institutions to reconsider "the underlying assumptions of modernization—the purpose of development, the measurements of progress and the use of industrial techniques."[55] As Barry Commoner, a leading American ecologist, put it, the "rescue rope offered to developing nations by modern science and technology was intrinsically unsound."[56] Early proponents of what would be called "sustainable development," like Commoner and Britain's Barbara Ward, urged global leaders to reevaluate the promise of industrialization through large-scale projects and argued that the terms of global development must be dramatically redrawn.[57] One example of this call to reassess development is the publication of *The Careless Technology* in 1972, a thousand-page volume of case studies compiled by seventy international scientists critiquing the ecological pitfalls of past and ongoing projects.[58]

Leaders in the Global South, on the other hand, were generally wary of this new shift in the terms of debate away from poverty alleviation and toward the "environment." From their vantage point, neither industrialization nor technology but underdevelopment threatened planetary health. They also feared a new rationale for international intervention into the management of national resources so soon after the end of colonial rule. Perhaps most famously, India's prime minister Indira Gandhi, at the United Nations' first conference on the global environment in 1972, argued that "the rich look askance at our continuing poverty" while at the same time "they warn us against their own methods."[59] Without jobs and the benefits of global technoscience, Gandhi argued, "we cannot prevent them [poor people] from combing the forest for food and livelihood; from poaching and from despoiling the vegetation. . . . The environment cannot be improved in conditions of poverty. Nor can poverty be eradicated without the use of science and technology." In Gandhi's conception of environmental destruction, poor communities and poor nations needed foreign aid in the form of science and technology if they were to lessen their reliance on subsistence exploitation of natural resources.

In the ensuing debates over the link between environmental concerns and development, the prospective recipients of "development" emerged as both victims of global environmental problems and the new perpetrators of impending global shortages. The narrative of fuelwood shortages captures this new double-bind. The choices of fuelwood users were limited by problematic development models, but a growing cadre of concerned observers who feared "spreading deforestation and desertification all over the world" quickly identified fuelwood users as "agents of destruction." As perpetrators and victims, they also became the "targets of campaigns to promote an 'environmental consciousness.'"[60] Concerns over receding forests led to a flurry of scientific studies and international interventions to identify, monitor, and intervene in the habits of local communities by teaching good resource management. These measures sought to save the rest of the world from the folly of poor communities misusing "global" resources. But within these targeted communities, their sustained reliance on firewood and charcoal did not reflect some sort of atavistic energy preference. Rather, fuelwood was always tied up with the rising price of oil and other "modern fuels" as well as the failed promises of modernization and the absence of alternative fuel infrastructures. They had little choice in the matter.

East Africa in the Global Environment

While concerns over wood scarcity implicated much of the Third World, in Africa the urgent narrative of resource exhaustion also fit in lockstep with an older narrative of "desertification," first written about by colonial scientists in the 1920s.[61] These scientists blamed encroaching deserts on intense human activity and their ideas traversed the continent, taken up by generations of researchers. In the following decades, the conclusions of these colonial scientists compelled sweeping colonial interventions into how communities farmed and used their land.[62] And while the precise causes of desertification remained debated and complex, the resulting imperial scientific literature both presumed and helped cement the idea that Africans were incapable of stewarding their own environments. Desiccated environments were even read as a reflection of Africans' heathen sinfulness, justifying Christian intervention.[63] These narratives of degradation coincided with efforts to also exclude Africans from land set aside as commercial forests or conservation areas under colonial rule.

East Africa's emergence into the crosshairs of the wood fuel campaign in the 1970s came about as a result of the prolonged drought that struck both sides of the continent for the better part of a decade. In these parched landscapes across the Sahel and down into Tanzania, experts surmised that the increasing scarcity of firewood would lead local communities to cut down trees and burn animal waste to cook their daily meals. Uprooting trees left loose topsoil at the mercy of prevailing winds, scattering crucial nutrients that were now not being replaced with manure. Over years, this caused desertification, deforestation, and the returning threat of famine. And while the effects of the Sahelian drought subsided by 1976 in many parts of Tanzania, it returned in the early 1980s. The cooking pot now sat at the center of what Northern environmentalists feared was a "vicious circle of deforestation, soil erosion, silting, uncontrolled flooding, desertification and other consequences of environmental degradation."[64]

In the next few years, a group of global environmental organizations sprang up, focusing their attentions on African environments. Founded after the Stockholm Conference, the United Nations Environment Programme (UNEP) opened its office in Nairobi in 1973 as the only UN agency dedicated solely to environmental issues and the only UN office located in the Southern Hemisphere. For environmentalists and policy experts in the Global North, the UNEP office was "a relatively small research institute that could carry out significant surveys and monitor environmental change."[65] Political leaders in Africa hoped that locating the UNEP in Kenya would have great political consequence in advocating for their own development interests. In 1977, the UNEP convened a conference on desertification in Nairobi where the increasing aridization of African farmlands fostered substantial political and scientific debate. After the conference, the UN passed a resolution to combat desertification's causes and suggested that all countries affected by it make fighting the expansion of the desert a top priority.[66] In these conversations, the process of desertification continued to be understood as predominantly man-made and caused largely by the subsistence practices of African peasants: fuelwood collection, grazing animals, and intensive farming.

Another new institution to take shape around these issues was World-watch, founded in the early 1970s and funded in part by the new UNEP. Worldwatch became the first global institute for environmental research. In 1975, it published its first major book, Erik Eckholm's *The Other Energy Crisis*, which crystallized the idea that global fuelwood shortages now posed a threat comparable and complementary to the oil crisis. Lester R. Brown, the founder of Worldwatch, wrote in the preface, "The scarcity of firewood directly affects that one third or so of mankind which uses wood as fuel. Indirectly it affects everyone, putting pressure on fossil fuel reserves and, as it diverts animal manure from fertilizer to fuel use, further aggravating the global shortage of food."[67] This was a global catastrophe. Eckholm argued that fuelwood shortages posed a "ecological threat to human well-being far more insidious and intractable than the industrial pollution of our air and water," and invoked the specter of the United States' Dust Bowl years as the price for the overconsumption of fuelwood.

With this increasing institutional focus on fuelwood, a flurry of local fuelwood studies became the first step in staging an intervention. And while a growing chorus of environmentalists were now critiquing international development's doctrine of faith in "science and technology," the key instrument for determining and quantifying the fuelwood crisis became these studies.[68] Subsequently, a variety of technological interventions emerged as the primary tools for intervening in and mediating the crisis. In fact, this research on fuelwood shortages reflected a new moment in which science took on a "central role in shaping what count[s] as environmental problems."[69] Computer modeling had quickly become a popular tool for defining the scale and scope of environmental problems after it was popularized by the Club of Rome's publication of *The Limits to Growth*.[70] The book's authors predicted resource shortages by using a methodology of systems dynamics (SD), which employed computer modeling to generate predictions of global exhaustion.

Utilizing similar methods, studies on fuelwood across the Third World relied on the "gap theory" to predict impending shortages. The "gap theory" argued (and modeled predictions) that due to rapid population growth, fuelwood demand would quickly consume global forests, without allowing enough time for regrowth.[71] In 1984 one research paper on Tanzania using the gap theory predicted that the "last tree would disappear under the cooking pot by 1990."[72] These forecasting efforts confirmed for many environmentalists that "the earth's finiteness" had become "increasingly self-evident."[73]

This kind of computer modeling and approach to the fuelwood crisis created what Peter Taylor has critiqued as an "undifferentiated moral and technocratic discourse."[74] The new massive scale of these "environmental problems" demanded an understanding of the biosphere as an interconnected and imperiled object. A problem conceived on this global scale required solutions on an equally massive scale; solutions could emerge only through a collective moral awakening that "everyone must change to avert catastrophe," which was of course hard to compel. The alternative was a technocratic fix implemented by "a superintending agency able to analyze the system as a whole can direct the changes needed."[75] In this new scale of the planetary problem, potential solutions were equally incumbent on massive global participation or the intervention of authoritative agencies to implement solutions.

Appropriate Technology

It seems no concerted effort was made to expand Tanzania's electrical grid or cultivate scaled access to alternative fuel infrastructures in order to address Tanzania's impending forest crisis. International lenders were pivoting away from the sorts of major infrastructural projects that had helped the Tanzanian state build three hydroelectric dams in the 1970s, which remained the primary sources of electricity for the nation. But even for the electricity generated by these dams, many obstacles to infrastructural delivery in both rural and urban areas stubbornly persisted. For example, villages sitting in the shadow of the Kidatu hydroelectric plant still used firewood as fuel.[76] Kidatu provided Dar with much of its electricity, but it was dependent not just on the amount of rainfall in any given season but also on the maintenance of a single-circuit 220-kilovolt transmission line that stretched over three hundred kilometers into the city.[77] This line, and thus the city's electrical supply, was vulnerable to disruptions from bushfires, animals, and the accumulating effects of corrosion. Fixing the line was also risky, as it traversed sections of Mikumi National Park where wild game could pose a danger to repair crews traveling rough roads in poor vehicles. Another source of power was the Ubungo station outside of Dar, which comprised eight diesel units and one gas turbine. By 1984, five of these diesel units were offline and the remaining three were in bad shape.

In the absence of major infrastructural projects, to intervene in a crisis of this scale actually required interventions on the smallest of scales. Change relied on convincing individuals and communities to adopt new technologies for cooking and compelling massive afforestation efforts.[78] State agencies and international donors proposed solutions that involved advocating for conservation, improving the efficiencies of biomass as fuel through the adoption of appropriate technology and community tree planting.[79] Presented as an alternative to the massive infrastructures and environmentally destructive uses of technology, appropriate technology was embraced by environmentalists in particular as key to a future where "small" would be "beautiful."[80] The Tanzanian state also avidly promoted appropriate technology as a potential escape route from dependencies on expensive Western technology and expertise. A workshop convened by Tanzania's Scientific

Research Council in 1977 optimistically predicted that villagers would soon be using "photovoltaic electricity generation, biogas generation, small-scale hydroelectric generation, and solar refrigeration and drying for food and/ or crop preservation."[81] By 1984 a World Bank report listed sixteen planned projects through the Appropriate Technology Project (AATP) alone, and the Swedish Development Agency had also opened an appropriate technology office in Arusha to keep track of and encourage projects.[82] And while in some instances these technologies were adopted, none came to replace Tanzanians' reliance on fuelwood.

At the heart of new attempts to create appropriate technology solutions lay efforts to reimagine the cooking stove.[83] A generation of more-efficient stoves, better charcoal kilns, solar cooking devices, and biogas generators began to pop up as Tanzanians were encouraged to reconsider what they could burn and how. This expanded repertoire included "forest residues, wood processing wastes, surplus softwood, agricultural residues, animal wastes and sugar industry residues."[84] Plans for a new factory took shape for fashioning wood waste into charcoal briquettes.[85] Meanwhile, a SIDO project funded by a Dutch university worked on how to use corncobs as a fuel for small gasifiers that could run village grinding mills.[86]

While animal wastes offered a potentially large store of energy when used in lieu of wood, repurposing dung sparked concerns over nitrogen loss if it were diverted from use as a fertilizer.[87] Animal waste could also be converted to biogas using biodigesters that transformed the waste into usable natural gases such as methane that could then power stoves. SIDO installed these biodigesters in the cattle-rich northern and central regions of Tanzania; namely, Mwanza, Arusha, Dodoma, Shinyanga, and Mara. There were about 120 biogas digesters installed by various projects across Tanzania and about 60 percent remained in working order by 1983.[88] To use the gas, though, people also needed to purchase new gas cookers (known as gobars). While some gobars were made in Tanzania, the plant that manufactured them relied on foreign steel, which was in short supply.[89]

Beyond the logistical problems of distributing, educating, and repairing this new profusion of appropriate technology, development officers in the field also faced the challenge of convincing people that these new devices would make their lives better. Or, rather, villagers still had the power to determine whether to use or dispense with such devices. Solar stoves could require women to stand in front of a giant vortex of heat to cook or else demand rearranging meals around times when the sun was still out. In an equatorial country where the sun set relatively early, this could pose a problem. One of the most successful models of a more efficient stove, the Kenya Ceramic Jiko, was designed particularly for urban users of charcoal. Kenyan women had helped design the stove, which could be readily made from clay and scrap metal scavenged from old oil drums. While it was supported and promoted by many aid organizations, ultimately its adoption and popularity reflected local word of mouth more than the zealous advocacy of foreigners.[90]

Despite the outsized hope that a technologically improved stove would be an easy panacea in the fight against deforestation, newer stoves were not always more efficient than old ones. Nor was there sufficient evidence to prove that traditional stoves were inefficient.[91] Technicians and aid workers also tended to forget all the other purposes a fire served beyond cooking meals: they warmed space, became a focal point of social gatherings, kept insects away, and could also dry food and provide light.[92] Just as trees and forests served multiple uses for local communities with fuelwood usually a byproduct of other interactions, fires also served multiple purposes. As Phil O'Keefe and John Soussan have pointed out, "the simultaneity of resource use is something that marks traditional modes of production and the disappearance of the simultaneity marks the transition to a modern commodity economy."[93] This loss of simultaneity rendered many of the stove options inferior for women, who, as the literature on fuelwood readily pointed out, were facing an expansion of their daily tasks and new encroachments on their time.

Beyond their potential pitfalls in utility, stoves as well as other new technologies faced the challenge of being exciting and aspirational. Much like living in a concrete house became an ingrained desire of modernity, so did electricity. The writer Binyavanga Wainaina humorously recalls the arrival of biogas collectors when he was a child in his community in Kenya:

> I was twelve years old, in a small public school in Nakuru.
>
> One day, the whole school was called out of class. Some very blond and very serious people from Sweden had arrived. We were led to the round patch of grass next to the parade ground in front of the school, where the flag was. Next to the flag were two giant drums of cow shit and metal pipes and other unfamiliar accessories. We stood around, heard some burping sounds, and behold, there was light.
>
> This is biogas, the Swedes told us. A fecal matyr [matter]. It looks like shit—it is shit—but it has given up its gas for you. With this new fuel you can light your bulbs and cook your food. You will become balanced dieted; if you are industrious perhaps you can run a small biogas-powered posho mill and engage in income-generating activities.
>
> We went back to class. Very excited. Heretofore our teachers had threatened us with straightforward visions of failure. Boys would end up shining shoes; girls would end up pregnant.
>
> Now there was a worse thing to be: a user of biogas.[94]

6.2. Newspaper caption: "Indiscriminate felling of trees is often the cause of desertification, which in turn compounds the hardships borne by the womenfolk who are forced to travel long distances to fetch water and firewood." Source: Daily News, July 3, 1985

International organizations promoting appropriate technology sought to create new solutions that did not mimic the large-scale development of the past and thus replicate its problems. As an afterthought, recipients of this technology were frequently asked to reimagine the material modernity of their own futures along with their daily routines. Using a new stove required overcoming a technical obstacle or rethinking how and when to make dinner, but it also required remaking a socially embedded practice.

Because both gathering fuelwood and cooking meals remained the work of women in East Africa and across much of the developing world, literature on the crisis also doubled as a burgeoning social science literature on how to improve the lives of women through international development. When researchers asked women about the new incursions on their time, they rendered the fuelwood crisis as not simply an environmental catastrophe but a sociological one. Just like late buses spurred liturgies of the wasted time of (mostly) male urban workers, stories of firewood shortages revealed the pressures on women's time. Perhaps the most repeated and evocative images of the fuelwood crisis that circulated in newspapers was some variation of women walking with bundles of firewood balanced on their heads or strapped to their backs. These women, unbeknownst to them, became the mascots of a new global cause.

In some cases, the stakes were about not just women's time but their apparent safety and security in their homes and communities as well. This played out in neighboring Kenya's response to the woodfuel crisis. A cartoon produced by the Kenya Woodfuel Program called *So, Firewood Can Wreck a Home?* opens with the image of a man chasing after his wife with a stick raised over his head.[95] The main character, Eshifindi, is described as a "rather foolish woman" slowly being driven mad by the lack of firewood available for cooking even though she lives next to a forest. "I can't touch any of them . . . or I'll be in for it! Everything is booked for either building with or selling!" she complains, demonstrating how states and communities gave other uses of wood priority over fuel, leaving women to compensate. Out of desperation, Eshifindi tears apart a chair in her home to fuel the fire for the evening meal. Meanwhile, a stranger comes to her village to sell her tree seedlings, cajoling her that the "fire must be lit": she has no choice but to find a way to cook food. "Come on, Mama! You are not that poor! Firewood is life! I'm Professor John—I know these things. I'm the pioneer of life." When her husband comes home, furious about the presence of another man and the broken chair, Eshifindi tries to argue it was broken "for . . . Hot water, Hot Porridge, Hot Ugali [cooked maize meal]." After Eshifindi's husband violently banishes her from the house, other women in the village get together and discuss how they also feel threatened by their husbands if they fail to make a hot meal at the end of the day. Ultimately, the story ends with Eshifindi's husband and another man from the village realizing that the answer to the fuel shortage is in "the management of trees."[96] Hardly a tale of female empowerment, the Woodfuel Program pamphlet promotes tree planting as a way to avoid domestic discord and violence. Women were apparently foolishly unable to find solutions while men shored up power and authority by turning to outside experts and the science of "managing" forests.

But women were not just the subject of such studies or the victims of deforestation—some also took up the cause. Most famously, Wangari Maathai became a prominent environmentalist and activist in Kenya, gaining attention particularly after the UN Conference on desertification in 1977 when she founded the Green Belt Movement (GBM). For Maathai, tree planting was both a cure for deforestation and a way to create jobs for Kenyans, particularly rural women. It was also an act of dissent from the state's policy of development that was privatizing public lands. Maathai's credentials as a scientist fashioned her as a particularly good ambassador for global environmentalism within Kenya. She also emerged amid the ecofeminist movement of the 1970s that drew on images of women as particularly well suited to intervene against the effects of patriarchal and capitalist abuses of nature. As Katherine Hunt has argued, Maathai's GBM sought to create an "empowering space for rural Kenyan women to enact their political consciousness."[97] By aiming the movement particularly at women, Maathai framed tree planting not so much as a salve for a global environmental problems as a path to self-sufficiency for a new generation of "women foresters" so they could more easily obtain many of their basic needs, "including wood fuel, building and fencing material, and soil conservation."[98] GBM has since become the quintessential example of successful tree-planting campaigns in Africa.

While no similarly vaunted figure emerged in Tanzania, by 1985 Kate Kamba, the secretary-general of the Tanzania Women's Organization (UWT), was in charge of spearheading new regional afforestation campaigns with women at the center.[99] Every district in the Dar es Salaam region was instructed to find areas that could be used for tree planting, where "small groups could open up tree plantations even within the proximity of their residences." Kamba reminded women in particular that it was their "role to step up food production as food was essentially a women's concern. Women were also urged to revive their traditional habit of growing flowers along their homes, offices, industries."[100] In an article published in Dar's *Daily News*, Eckholm of Worldwatch also called for women foresters to work with rural women, as they were the primary collectors of fuelwood. Fuelwood studies also warned that because collecting fuelwood was considered "women's work," men were slow to help in some communities.[101] Defining the crisis of global deforestation so particularly as a consequence of fuelwood put subsistence practices and women at the center of both the problem and solution; it also tacitly exempted commercial uses of forests as not the cause of destruction. Whether they embraced it or not, women found themselves characterized by the fuelwood crisis as particularly vulnerable and ecologically minded.[102] And, for all the discussion among development experts of how deforestation was infringing on women's time in the new profusion of studies, women were now expected to fill yet another role in social reproduction: planting trees.

Planting Trees

While research continued into new ways to glean energy from biomass and new versions of the cooking stove abounded, international afforestation programs like those helmed by Maathai and Kamba became the most concerted intervention into the fuelwood crisis. These efforts were animated by an onslaught of alarming statistics predicting global forest loss. In the Food and Agriculture Organization of the United Nations (FAO)'s 1977 annual quantification of Africa's forest resources, they predicted that "a large part of . . . unreserved wooded areas will disappear. These unreserved forests and open woodlands supplied much of Africa's saw timber" as well as fuel.[103] Overall, 75 percent of Tanzania's closed forests were already reserved (set aside either for protection or for formal timber exploitation), and thirty percent of its woodlands. By 1982 the FAO was estimating that an average of 11.3 million hectares would be destroyed per year in the next four years as communities took to unreserved forests to meet their fuel needs.[104]

6.3. Planting seedlings. Photo courtesy of Tanzania Information Services

A decade's worth of dire predictions inaugurated a flurry of tree-planting projects across the globe. In Tanzania the state launched a series of public awareness campaigns funded by foreign donors focusing on regions where fuel shortages were most acute. These early afforestation projects focused on recruiting ujamaa villages to start community woodlots in order to meet their fuelwood needs and keep them out of neighboring forests.[105] The state then launched a more aggressive tree-planting campaign in 1980 called Misitu ni Mali (Forests Are Our Wealth), followed by an additional village afforestation plan in 1982, and another effort in 1984. Over the course of Misitu ni Mali, Tanzania's Forestry Division aimed to drive home the importance of tree planting through adult education campaigns, distributing half a million technical pamphlets in Kiswahili across seven of the driest regions of the country and making educational radio programs including a radio play about the importance of tree planting.[106] They also organized seminars and discussion groups and film screenings, made T-shirts and posters, and enrolled the assistance of churches, training institutions, and over 1,500 primary schools, 49 secondary schools, and 30 teacher training centers.[107]

The Forestry Division reported receiving over eight hundred letters from farmers in response to the campaign in the following six months, suggesting it was effective in drawing attention to tree planting.[108] Other reports, however, critiqued the slapdash planning of Misitu ni Mali, since regional forestry staff had not even been consulted. The pamphlets, for example, only discussed planting pine trees, which were not appropriate in certain regions. And while the campaign was promoted over the airwaves, a chronic battery shortage by the 1980s left many village radios silent.[109] A Peace Corps volunteer, writing in the *Arid Lands Newsletter*, noted that eighty thousand seedlings had been brought to the district where he was volunteering and "simply dropped off at the villages and schools with instructions to 'Plant them in your woodlots.'"[110] He pointed out that villagers did not actually see a need for fuelwood trees but would have been very happy for fruit trees.

By one estimate, in just two years (1984–1985) eighteen million trees were planted in nineteen different regions.[111] And yet anxieties over fuelwood did not seem to subside, despite these numerous projects. Reports to donor agencies frequently portrayed communities as part of the problem. Farmers were "not people to be relied on" in solving the fuelwood crisis. They were irrational users of wood in general and were reluctant to plant trees; "securing the cooperation of the local people" presented "formidable problems."[112] An FAO report in 1990 looking back at attempts to intervene in the crisis argued that "the rural people themselves did not perceive the problem then and have not done so to date."[113] Tanzania's director of forestry, E. M. Mnzava, bemoaned that it was not a matter of how many hectares were planted but rather how aware people were of the problems of deforestation and how willing they were to "take any steps."[114] And yet this inability of local communities to perceive the problem did not readily lead researchers to question whether the nature of the problem itself needed to be reassessed.

One key point of disconnect seems to have been the fact that the majority of charcoal and firewood production came from trees outside of woodlots and protected areas. Plantations and woodlots played only a negligible role in supplying communities with fuelwood because they tended to be prioritized for other uses such as poles for building.[115] Researchers found it frustrating that communities were reluctant to plant trees for fuel and yet they did not instead provide fruit-bearing trees that could serve multiple uses in their lifetime before ending up as charcoal. It seems that for all the focus on afforestation and fuelwood, many researchers seemingly missed the reality of how communities appraised forests and used trees. Or rather, Tanzanian foresters and international researchers did not miss this reality so much as they hoped to fundamentally change it.

In conjunction with these local interventions, Tanzania's press published a steady stream of articles on the fuelwood crisis as well as on the crisis of desertification. Some of these articles were written by local experts, but many were by Earthscan and Worldwatch researchers who aggressively sought out local coverage of the fuelwood crisis in newspapers across the world.[116] Across East Africa a steady stream of campaigns managed to bring the environmental movement and conservation efforts into an immensely expanded public discourse in just a few years.[117]

But like new cookstoves, the promise of tree planting as a transformative technology could only be realized through its widespread adoption by local communities. Despite appropriate technology's romantic slogan valorizing the small, afforestation and the adoption of new stoves had to be big to be beautiful; as technological fixes, both depended entirely on the vagaries of their scale. One new stove or one enthusiastic villager planting trees would never solve a regional or global crisis. Because of the need for mass involvement, both state and international tree-planting campaigns fought against the whims of local participation by invoking a language of common sense.

Planting trees has also always been ripe for analogy and metaphor. Across diverse cultural contexts, trees symbolize qualities of rootedness, "possibility, growth, and change" and as a result they have become weighty with inferred meaning.[118] In the twentieth century, the act of tree planting became imbued with an air of "common cause" that yielded unmitigated benefit.[119] Because it evokes community stewardship of resources and requires no specialized technological knowledge, someone reluctant to plant trees became correspondingly irrational, self-defeating, and environmentally irresponsible. Within American history, tree-planting campaigns are frequently narrated as both mythical and mundane acts of unmitigated good, capturing civic consciousness and environmental stewardship.[120] The relevance of American history here is not merely to serve as a point of comparison. In 1972, at nearly the same moment that Northern environmentalists began sounding the alarm over fuelwood, organizations such as the National Arbor Day Foundation came into being in the United States to promote tree planting across the continent. Like the seedlings themselves, tree planting soon traveled the globe as the universal answer to a problem equally expansive in scope.

But in radiating out from the West to the Global South, these tree-planting campaigns also mirrored a darker colonial history of arborescent culture. The benign insistence on the goodness of trees has long served to settle and domesticate lands, cover over violence and theft, and reinvent environments in the bucolic image of colonial occupiers.[121] And yet despite these imperialistic undertones, the Tanzanian state latched on to the metaphoric and real potential of tree-planting campaigns for their own purposes. Arborescent imagery and metaphors have historically also been used to portray communities and nations as "rooted" in place, ancient and natural.[122] Indeed, the Tanzanian state's embrace of tree planting was shaped not merely by a sense of ecological urgency but by their marked effort by the early 1980s to forge new futures, independent from past reliances on foreign oil. The state sought a future rooted in Tanzanian resource sovereignty.

Sustainable Development

Both international organizations and the Tanzanian state took up the specter of deforestation to compel their own contrasting versions of the nation's
future. The identification of a crisis always invites moments for recharting
the future. For international actors, the salve was sustainable development.
Along with other imperiled aspects of the global environment, fears over fuelwood together with the devastating economic effects of the oil crisis helped
usher in the new era of externally imposed "sustainable development." On the
other hand, seeking "groundedness" is what I call the socialist state's strategy
in the early 1980s for confronting the accumulating debt and increasingly
constricted terms of international aid.

The constant calculation and prediction of forest loss may have reflected
a crisis foretold, but it was also its very construction and antidote. As defined
in the UN's Brundtland Report in 1987, sustainable development is "development that meets the needs of the present without compromising the ability
of the future generations to meet their own needs." Sustainable development
was born of crisis and its imposition came through identifying instability,
emergency, and dysfunction as the new normal in African life that needed
to be averted. As the media, development experts, and multinational lending
agencies created the expectation of ongoing exceptional circumstances, a
certain set of international actors, particularly nongovernmental organizations (NGOs), were empowered to define a new "common sense" and prescribe solutions.[123]

In the case of the fuelwood crisis this led foreign organizations to urge
local, "sustainable" solutions such as tree planting while undermining the
notion that local communities or even national governments were the best
stewards of the "environment" or "development." Since the late 1960s many
Northern environmentalists had come to see African states as reluctant to
intervene on behalf of their attempts at wildlife conservation. In East Africa,
Stephen Macekura writes, leaders within the American and British conservation movements grew impatient with their thwarted attempts to convince
nationalist leaders to set aside conservation land for wildlife. In their frustration, they shifted their tactics in the 1970s and turned their attention to
countries providing development assistance.[124] At the same time, international lenders grew critical of African states as corrupt, inefficient managers
of development money and begun directing funds to a growing number of
intermediary NGOs.[125]

By the early 1980s there was unprecedented global attention on African
environments as well as a growing consensus that sustainable development
could reform the mismanagement of African states and natures. From 1980
to 1990, over 1 billion dollars were funneled into forestry projects in Afri-

ca.[126] Yet as communities were conscripted into tree-planting efforts by dire predictions of deforestation, an accompanying wave of neoliberal reforms privatized forests and created new agendas along the dictates of conservation, biodiversity, and tourism.[127] Newly powerful NGOs lobbied international agencies and banks to consider conservation as a condition of development assistance.

By 1985, when Nyerere announced that he would step down from the presidency the following year, he had spent the previous five years battling both internal and external pressures to institute dramatic economic reforms. These reforms, which became conditionalities of borrowing from the IMF, included liberalizing Tanzania's markets and opening land to foreign investors. The IMF also insisted on devaluing the Tanzanian shilling, which drove up the price of foreign fuel. While Nyerere held out in the early 1980s against these conditionalities (agreeing instead to implement domestically determined austerity measures), Tanzania lost 25 percent of its foreign aid as the IMF and World Bank pressured formerly friendly governments to compel Tanzania into compliance.[128]

Ironically, many of these policies only made urbanites more reliant on charcoal and exploiting local forests as urban infrastructures remained underfunded and urban prices high. These policies also brought more foreigners into the forests. Once the IMF prevailed in devaluing the Tanzanian shilling, a new wave of tourism in game reserves and national parks began raising the profile of conservation as a key contributor to the national economy.[129] Exports from Tanzania's forests increased tenfold, from USD 1.4 million to over USD 11.6 million in the next five years.[130] Through expanding conservation, tourism, and forestry, Tanzania's trees became monetized for foreign consumption at a time when local communities were urged to stay out of forests and plant trees in order to allay their own reliance on such resources. As Thaddeus Sunseri has argued, Tanzania's forest policy now reflected "the hegemony of biological preservationists and the priorities of bi-lateral and multilateral donors and international and local NGOS."[131]

Getting Grounded

While the state was certainly at odds with many of these new reforms, an ongoing national call to plant trees—with Nyerere himself frequently announcing such efforts—echoed and amplified the urgency of international environmentalists. In 1981 Tanzanian foresters argued that everyone would have to plant thirteen trees per year in order to achieve a balance between planting and deforestation, seventeen times more trees than had been planted that year.[132] "We are speaking of survival when we talk about afforestation," Nyerere remarked a few years later when announcing Misitu ni Mali.[133] Between 1970 and 1984, two hundred thousand hectares of trees were planted through

local efforts largely funded by foreign donors.[134] Despite the ultimate effects
many of these foreign interventions had on Tanzanian sovereignty, Third
World leaders like Nyerere had come to see some international organizations
and their potential to function as regulators of commodities and capital as the
last bulwark between themselves and remaining under the old economic order. In the hopes of limiting this "loss of control" that so many countries were
facing, Third World nations turned to international organizations, hoping
they might channel and consolidate their power and desires.[135] Countries like
Tanzania sought economic independence somewhere in the rocky terrain of
national resource sovereignty, international aid, and the ongoing experiment
of Third World solidarity first started at Bandung in 1955.[136]

But while the Tanzanian state embraced the aid and tactics of international environmentalists to counter the fuelwood crisis and desertification,
I would argue that it hoped these campaigns could enable a different future
than those imagined by Western environmental interests. Tree planting and
appropriate technology were not simply tools of the new vanguard of externally imposed sustainability; they also had the potential to reenergize the
attenuating solidarities of the Third World. Nyerere anticipated these sorts of
activities would achieve a collective bargaining power for the Global South.
Tanzania could lessen its dependence on foreign oil through better management of its natural resources. In this vision of the future, afforestation efforts
and a new focus on the environment could restore sovereignty to national
development through provisioning the state for its own needs and weaning
the economy from the vicissitudes of foreign exchange.

In the years following the Arab oil embargo, Tanzanian development
had failed to be realized in the material terms first imagined. In the shadow
of rising fuel prices, dreams of concrete were rewritten into a state ideology
of burnt bricks as a vernacular Tanzanian building alternative invulnerable
to fluctuating oil prices. By limiting private car imports and prioritizing city
buses, oil shortages made what was economical also environmental. Factories were forced to reconsider the value of waste and explore overlooked local
resources. Due to the ongoing reality of shortages, Tanzania was in many
ways well on its way to experimenting with what the West might call "sustainable development." But these material transformations did not emerge
from a conservation agenda so much as from an experimental and sometimes
desperate need for materials mixed with the politics of economic revolt from
the old order. Facing this renegotiation of how the future would literally ma-
terialize, Nyerere and many other leaders of the Third World saw the oil crisis
as a lesson in economic sovereignty. Greg Grandin has recently written that
in the 1970s a "carbon solidarity" emerged among weaker nations hoping to
use "oil as leverage against the strong."[137] The Organization of the Petroleum
Exporting Countries' move to increase their own bargaining power against

the industrialized world both harmed Tanzania's economy and served as a powerful example of what was possible in shaping a new economic order.[138]

During his struggles with the IMF, Nyerere received the Third World Award, a prize honoring his contribution to development in the Global South. In front of forty-four other national leaders in New Delhi, where he received the award, Nyerere declared with urgency that the assembled countries must rethink their models and modes of development. Echoing Northern environmentalists like Commoner and Ward, he argued that the North was peddling a kind of development that both harmed the South and was proving to be globally unsustainable. For him, though, the answer was for the Third World to turn inward, to "our own roots and our own resources," invoking a metaphor that captured an evolving language of "sustainable development" peddled by the very institutions that would erode sovereign decision making over natural resources. Indeed, despite Nyerere having been invited by Indira Gandhi herself, his critique of Northern technology and science as saviors of the Third World was in stark contrast to Gandhi's embrace of foreign technology and international development as a panacea for poverty, ten years earlier in Stockholm.

"The Third World, in its relation with the North," Nyerere suggested to the assembled audience, "is like a trade union in its relations with employers."[139] The Third World had little choice but to band together against the growing demands and conditionalities of Northern "institutions of world economic management." These were ideas Nyerere had been at pains to promote among the Group of 77 and the nonalignment movement; only in banding together and creating such mechanisms and institutions such as "Third World Multinational Corporations" could the Global South acquire the necessary weight to rewrite the terms of global development.[140]

In a similar vein, Amir Jamal, Tanzania's former minister of finance, two years later at another meeting in India advocated for a common Third World market as a path to collective sovereignty. Jamal asked the audience to "see with the eyes of a caricaturist": "If the world is a see-saw fulcrummed on a North-South axis, what fun is it for the pot-bellied North to find itself suddenly bumped, to be grounded by its own weight, simply because the feather-weight sitting on the South end of it failed to provide it with the essential counterweight, having got itself flung into the air in a flurry of so many straws swept by the wind of the North?" Jamal went on to suggest that developing a common market of materials as well as intellectual collaborations could provide this necessary counterbalance. The Global South needed a turn on the ground, leaving the corpulent North, for once, hanging in the wind.[141]

In this task of renegotiating the future, the charcoal economy that had expanded to fill the absence of "modern fuels" in Dar offers perhaps the best example of a rival form of sustainability even while portrayed within the narrative of the fuelwood crisis as the apocalyptic, all-consuming opposite. From this vantage point, I would suggest that state afforestation efforts in Tanzania were only partly interested in the same things as Northern environmentalists. Before the Sahelian drought in the 1970s and the threat of global deforestation, the state had clearly begun to test the waters of an international charcoal trade. Afterward, as one study critically suggests, the Tanzanian state "may have realized that promoting afforestation is more profitable than halting deforestation."[142] Without any ready energy alternatives, planting trees was always an act of fuel creation as much as it was an environmental aspiration. The Forestry Department still welcomed help to fund afforestation efforts—and by 1984, 95 percent of the Forest Directorate's budget was funded by foreign assistance—but there was seemingly no parallel intervention into the charcoal trade.[143] Instead, the state sought to expand the ways in which trees could play a more prominent role in domestic industries. Factories like the Mufindi Pulp and Paper Mill, which at the time ran on coal and petroleum, were slated to be modified to instead consume fuelwood and aid in cheaper production.[144] On Tanzania's smaller islands, where electricity was hard to access, projects were initiated to also glean fuel from softwood plantations and coconut husks. In 1975 the Tanzanian finance minister, Cleopa Msuya, explicitly urged the Tanzanian state to cut the high costs of energy importation by focusing on the "production of hydro-electricity . . . and use of charcoal"[145] Expanding charcoal—along with other alternative energy sources—provided a potential escape route from the stranglehold of foreign oil.

This new exploration of ways to use charcoal and trees might have echoed the international movement toward appropriate technology and reforming development, but it ultimately gave shape to a different notion of "sustainability." The early 1980s in Dar were dotted with conferences and visiting experts that argued for abandoning Western modes of development for domestic alternatives. In 1983, for example, Gerhard Kohler, a technocratic expert from East Germany, visited the university to lead a one-week seminar with the heads of planning for various ministries to develop a plan for cutting down on foreign imports. Kohler told the *Daily News* that "the approach should be to try as much as possible to be inward looking by exploiting local resources instead of seeking foreign assistance." As a point of departure, he urged Tanzanians to see East Germany as an example of looking inward as a way to overcome these economic problems.[146] In a science symposium at the university in 1985 a Dar professor also urged Tanzanian technocrats to guard against "technological invasion" that would only encourage "the wholesale adoption of foreign technologies" despite the fact that it had brought "adverse effects on society and environment"[147] The minister for planning and economic affairs meanwhile announced ongoing projects to extract coal, kaolin, soda ash, and phosphates from the coastal region in order to "change the present economic order." These narratives, similar to the critiques of Northern environmentalists, saw the environmental crisis as a legacy of Western development, but it did not likewise locate the solution in the same doctrines of conservation but rather in resource sovereignty. Tanzania reduced its oil consumption by 20 percent in only one year out of necessity rather than an ecological imperative. This clearly had devastating effects on the state's economic capacity but it also served as a motivation to seek alternatives. The ongoing fiscal crisis no doubt sparked a broad reassessment of resources and relationships.[148]

While both the state and local communities were turning to charcoal economies, this did not, however, mean that their interests aligned in local forests. Whereas charcoal production had in the early 1970s mostly occurred in a fifty-mile radius around the city, rural families as far as two hundred kilometers away became part of the urban charcoal economy in the next decade.[149] In 1979 cashew farmers in the Rufiji river valley (a few hours by car from Dar's charcoal-using urbanites) were facing such disastrously low global market prices that farmers were giving up on their cashew trees and turning to charcoal production instead.[150] Government officials were frustrated by this new development, since in general "cashew nuts bring foreign exchange, charcoal does not."[151] To stop farmers from forsaking their cashew crops or turning them into charcoal, the state enacted a bylaw that banned locals from fishing or making charcoal, forcing them to tend their trees instead.[152] Farmers were also prohibited from leaving their region without first proving they had weeded their cashew crops. Regional authorities closed local markets to discourage other forms of trading and the local party secretary in Kibiti arrested ninety farmers and took them to court for neglecting their cashew trees. Those unable to pay their court fees spent up to three months in jail.[153] Despite these measures, cashew nut production still faltered while charcoal production continued. While only 3,521 bags of charcoal were sold in June (compared to 24,054 bags the month before), locals seem to have simply waited until the end of the ban to sell it again openly. In fact, the ban on charcoal production may have ultimately been a boon to local producers and transporters because it created an artificial shortage and prices skyrocketed.[154] This shift from cashews to charcoal mirrors what also happened among coffee farmers in the Kilimanjaro area who began cutting down their coffee bushes to grow tomatoes and onions when the bottom fell out of global coffee prices.[155] While the state might have been interested in forging a charcoal economy and weaning the nation from a dependence on foreign oil, it did not seek this at the cost of their foreign-exchange-generating exports.

It is clear from the clashes in Rufiji that for local producers charcoal was a way to survive the rupturing economy whether the state sanctioned it or not. Urbanites and their peri-urban producers were seeking their own version of groundedness and economic sovereignty that evaded both the state and international imperatives. These Tanzanians crafted an economy in the absence of oil and foreign exchange with local forest resources at the center. Like the cashew farmers along the Rufiji river, urbanites were not driven by state desires to create foreign exchange when more money could be made in other ways. They were, however, encouraged by pervasive shortages to avoid relying on products that required foreign exchange. Just as a bag of charcoal took very little foreign input for locals to produce and bring into Dar for sale, it also did not require any to use, beyond acquiring a basic stove (*jiko*), frequently welded by small industry craftsmen out of salvaged metal over a hot charcoal stove.

Many of Dar's residents also simply preferred the taste of food cooked over charcoal, and the resulting cuisine was a predominant feature of the "informal" economy by the 1980s. Women, especially, set up charcoal stoves in the interstices of the city where they sold roast corn, fried donuts known as maandazi, and chapati bread. By the late 1970s hundreds of women made the trip every day from their homes to the fish market in downtown to then return and fry in the evenings along the road on charcoal stoves, lit by the small kerosene lamps known as *kibatari*.[156] In an illustration of how fuel sources sometimes collaborated, these kerosene lamps were made of scrap metal and molded, shaped, and welded with a "makeshift welding rod heated in a charcoal stove."

Coffee sellers also walked the streets or camped out under trees with vessels (*mdila*) of coffee warmed by small charcoal-filled canisters attached to the bottom for mobility.[157] Perhaps skewers of grilled meat known as *mshikaki* most aptly capture how street food both literally was fueled by and fueled an economy in the shadow of foreign exchange. A favorite of Dar residents, mshikaki was grilled over a charcoal fire using old bicycle spokes as kebab skewers. Diners had to return the spokes after eating. Charcoal also fueled multiple small industries that women initiated to make extra money. The UWT started projects in beer brewing and pottery that relied heavily on fuelwood and charcoal.[158] Charcoal was used to dry tea leaves and smoke fish.[159] There was no shortage of ways that charcoal might compensate for other fuel shortages across the constellation of new informal and formal work in Tanzania. In many ways, Dar's urban residents were far more successful than the state in establishing an independent economy out of the surrounding forests. But this is not to be overly triumphant about an economy forged out of necessity.

Despite state efforts to forge a new Third World economic order, "sustainable development" emerged out of the 1980s as the new "common sense" lexicon of progress. But the end of the fuelwood crisis is punctuated by two important ironies that belie any discursive tidiness. The first irony returns briefly to the nature of the fuelwood crisis itself and how it emerged. While there is no doubt that charcoal production can leave behind a problematic environmental footprint that Tanzanians still struggle with today, the pronouncement of crisis dramatically missed its apocalyptic mark.[160] We do not live in a deforested world, nor has Tanzania been denuded of trees as was predicted. Scholars looking back at the fuelwood crisis have since seriously challenged the view that firewood collection in particular was the leading cause of deforestation.[161] In the case of Tanzania, deforestation in the 1970s was a product of a host of factors, and mainly driven by state-sponsored villagization and land clearing for farming.[162] The hypothesis that fuelwood collection and use was the predominant cause of desertification and contributed to the effects of the prolonged Sahelian drought has also been revised. In fact, the most recent theory of what caused Africa's prolonged droughts in the 1970s and 1980s suggests that it may have been the very industrial pollution that Eckholm had posited as a lesser threat to human survival than the fuelwood crisis. New research on the effects of pollution in the 1970s suggests that the Sahelian drought may have been caused by aerosol particulates accumulating in the atmosphere above the Northern Hemisphere. As these particulates accumulated, they produced cooled air that shifted the "intertropical convergence zone" from its usual position along the equator of Africa, where it brings crucial seasonal rain. Instead, these rains fell further south.[163] In their absence, deserts and cities grew.

Secondly, the profound importance of the charcoal economy continues today as no affordable alternative has been cultivated. It also remains an anarchic fuel. In a 2009 study of the industry, the World Bank estimated that charcoal provided "income to several hundred thousand people" in both urban and rural areas and generated over ten times the revenue of both coffee and tea exports.[164] Small-scale charcoal production and the people who sell and distribute it remain the fundamental energy infrastructure of one of the fastest-growing cities in the world. This remains true in many African cities today.

Conclusion

PROVISIONING FOR AN UNKNOWN FUTURE

At the risk of reducing the preceding chapters to a litany of small and large calamities that shaped Dar's environment, I want to end by writing about failure in African history. By 1974, the Tanzanian state had begun the process of compulsorily relocating its rural population into ujamaa villages as well as relocating its administrative capital. These monumental endeavors (to just name two, of many) then unfolded among an extended drought, the collapse of food production and distribution, and the first effects of the oil crisis. Needless to say, things did not go as planned. Sparked by this host of larger forces, new and old residents of Dar transformed their city as they sought to cope and sometimes prosper amid these swiftly changing circumstances. In narrating this history in the decades since, policy experts, economists, and development practitioners—many who were part of this history—have tended to consider this period in Tanzania as one of utter crisis.[1] In response, historians like Priya Lal have rightfully called out the cynicism with which ujamaa has been glossed as an outright, almost foretold, failure.[2]

Lal's call to avoid this cynicism echoes a chorus of scholars of the continent who have also begun critiquing how much salience and power the media, states, and experts have given the notion of "crisis" to explain everyday life in Africa—sometimes to dramatic and violent effect. Brian Goldstone and Juan Obarrio point out that the term has become "a structuring device that, far from simply appraising the quality of this or that phenomenon vis-à-vis a particular calculus or within a specific narrative, literally constructs the narrative itself."[3] In diagnosing a crisis, what is inherently multicausal and complicated is shorn into simplistic and bold terms as a way to both assign blame and shore up power to prescribe (and dictate the terms of) a solution.[4]

Crisis as narrative structure is not particularly new; in the past half century, it has routinely shaped the conditions of possibility for independent Africa as part and parcel of development.[5] The global development paradigm of the 1960s and 1970s was based on the binary of stability and crisis, with the latter kept at bay through the authority of foreign expertise. The social sciences and the states and institutions that enrolled these forms of authority in their own postcolonial projects, notes Timothy Mitchell, have given immense power to "expertise, reason, explanation, and simplification."[6] There was not much room for the "complex and indeterminate" in these models of intervention, and where indeterminacy did exist it served to justify more expertise, more intervention.[7]

Debating the relevance of the expert's approach to development also shaped postcolonial Tanzanian politics as they were unfolding. For example, by the 1970s several faculty members at the University of Dar es Salaam became reluctant to cooperate on state development plans because, as expatriate faculty member Andrew Coulson remembers, the state preferred instead to "use foreign consultants. Many university staff and students were openly critical of management agreements with multinational corporations."[8] Their skepticism of the foreign expert mirrored an ensuing conversation over whether the kind of disciplinary thinking fostered at the university was helping or harming Tanzania's future. These conversations led to an attempt to overhaul a vast majority of academic programs in order to reformulate a curriculum to better address broad development goals. It was, in no small part, an attempt to decolonize higher education.

Proposed by a group of leftist faculty who called themselves the Group of Nine, the rationale for this new curricular arrangement was their collective belief that the "increasing specialization of knowledge, and the constitution of these 'specialties' into self-sufficient academic disciplines each with their own 'laws,' theory and techniques, can be related to the historical development of capitalism."[9] They sought to avoid draining Tanzania's intellectual resources with these practices and instead transform the meaning of "specialized or technical knowledge."[10] Their goal was to cultivate students who would avoid falling into the trap of the "specialist" or expert when there was so much "development" to be done.[11] They cautioned against creating a new vanguard of intellectual elite too specialized to think broadly about the cause and needs of Tanzanian socialism. And yet, while framed as an opportunity to rethink higher education for Tanzania's needs, the new plan was ultimately voted down by a growing contingent of Tanzanian faculty and in particular a history faculty who were wary of losing their place in this new curriculum.[12] These Tanzanian scholars were returning with foreign PhDs and perceived this new approach as a "real challenge to the very foundations of their own hard-earned expertise."[13] Clearly, the tension between the expert and the generalist, the crisis and the complex conditions of development, was never far from the conversation about Tanzania's future.

While the preceding chapters consider the effects of the "oil crisis," the "woodfuel crisis," and the "urban crisis," I have explicitly not focused on the disciplinary thinking of the expert nor aimed to corroborate the narrative device of crisis. Indeed, the all-encompassing kind of catastrophe, sewn up neatly in such terminology, is quite the opposite of the kind of failure this book considers. If anything, these chapters seek to be a corrective to the simplifying language of crisis while still describing the reality of breakdown and chronic shortages that Tanzanians faced. As urbanites found ways to make livelihoods in tough circumstances, they also made new urban landscapes. In looking at the negotiations over building materials, transportation, fuel, food, and waste that transpired over the course of the 1970s, my focus remains on how contingent and vulnerable the daily workings and failings of the postcolonial city were. Residents navigated these brittle realities by finding ways to make their daily lives and environments malleable: molding infrastructures and landscapes to accommodate their changing circumstances. Indeed, sometimes the failure of infrastructural systems was both what curtailed certain opportunities and what brought about new ones. State bureaucrats, factory managers, and state-affiliated university research institutes at times enforced the unfolding austerity of the city, policing lines of urban belonging, profiting off shortages, and defining the culprits of crisis. But at other times they were themselves deeply committed to reimagining how things could be done while also facing the necessity of reinvention in their own lives.

Writing about how urban lives and environments were shaped by overflowing cesspits, unpaved roads, and food rationing could easily be read as a narrative of Afropessimism. And for a continent that has so often occupied the slot of "failure" in civilizational narratives, it is certainly problematic to unthinkingly abide by such terminology. Urban theorists who write about the continent's cities have grappled with a version of this question as well as who is implicated in narratives of failure. Should the "dysfunction" of the continent's cities (when read as an implicit comparison to cities in the Global North) be written as a reinvention of urban forms, or is it critical to name these as dysfunctions in order to highlight the dire circumstances millions of Africans face daily as well as the culpability of African states and global capitalism?[14]

Instead of agonizing over how to gloss these urban forms, I suggest that we reconsider how we render failure in African history while remaining alert to the ugly historical legacy and ongoing potency of such language. In this regard, I am interested in failure as a process, not as a diagnostic. Rather than conceding that failure is part and parcel of the persistent narrative of crisis that lurks in so many stories of the continent, perhaps it can be rescued as a bulwark against such intractable narratives that abstract and scale up. Jack Halberstam writes, "Under certain circumstances failing, losing, forgetting, unmaking, undoing, unbecoming, not knowing may in fact offer more creative, more cooperative, more surprising ways of being in the world."[15] Failure can be fertile, imaginative, and experimental, and it can suggest the possibility of new worlds while also recalibrating past expectations of the future. It might also serve as a way to interweave material and intellectual histories of place: how did certain quotidian failures reshape the materiality of the city as well as larger state ideas about resource use? It might be a way to stay close to the ground in how we tell stories about the past.

Writing about failure is also essential historical practice. Richard White argues in his book *Railroaded* that if historians do not write about failure, "all that is left for historians to do is explain the inevitability of the present. The inevitability of the present violates the contingency of the past, which involves alternative choices and outcomes that could have produced alternative presents."[16] Many possible futures and unfolding presents are captured in this book. It need not be a slander to acknowledge that some situations were unintended and unwanted. Crisis, on the other hand, tends to paralyze the narrative, invite the justification for external intervention, and explain the inevitability of a certain future.

By approaching Dar es Salaam's landscape with a focus on the multicausal and cascading problems that marked urban life, perhaps we can move away from the tidy packaging of crisis and the rise of the expert that has become central in colonial and postcolonial historiographies. Instead, if our attention remains with the disruptions faced on the ground, we are forced to stick to the everyday and how solutions born of necessity far more than urban master plans came to shape people's urban environments and futures. It demands a focus on what Michael Degani has recently called the state of "permanent improvisation in urban Tanzania."[17] In little more than a decade, Dar became a city redefined from its outsides in, as new migrants arrived on the city's edge and the center migrated to the periphery in order to provision their lives. In the face of ever-changing circumstances, the city became more rather than less entrenched in "nature," reshaped by new flows of people, food, and waste as state bureaucrats and citizens jostled in the urban margins and in print over who would define productivity in the city. Urban authorities sometimes dreaded the flexibility of people to take on new jobs, access new resources, and travel new urban itineraries, and yet mastering changing circumstances also became an expectation of urban citizenship.

In considering the "permanent improvisation" that is required in daily life, there is a risk of remaining only on the surface. It is crucial to note that beneath the urgent creativity of remaking landscapes and materials, there were also shifting ideologies about Tanzania's future. Sometimes these ideologies were what sparked material transformations and other times they were only articulated in the aftermath of these reckonings, as the state hoped to retrofit dramatically changing circumstances into their own narrative. If White might urge historians that considering failure is essential for leaving the future open and not inevitable, Dan Magaziner, in *The Art of Life in South Africa*, cautions against the penchant for historians to focus too much on the future rather than the imperfect—perhaps even crass—negotiations of the present. In writing about a South African art school that produced art teachers for the apartheid system rather than the famous artists that tend to occupy the historiography, Magaziner observes that "African intellectual history" is far more comfortable writing "histories of the future, not explications of a series of presents." Like failure, writing in the subjunctive about what should have been or what one day will be is crucial for revealing the possibilities of any historical moment. It also reveals the capacious visions of historical actors. But it might overlook those who reckon with the daily trials of the moment and, consequently, the complexity of historical actors who face constantly shifting horizons.

If anything, this book narrates the persistent need of both the state and citizens to recalibrate and quite literally reterritorialize the materials necessary for daily life: Where can we now find fuel? What do we replace cement with? Can factories also supply their own raw materials? These present concerns also mean recalibrating the material possibilities and politics of the future. Thus, they are intellectual endeavors as well as improvisational ones. Charcoal, for example, was not just the forgivable last resort of the desperate urbanite but an alternative development paradigm that opened the possibility of a different present and future.

Engagements with the material reckonings of nations and communities have been somewhat absent from the postcolonial story, siloed instead into a literature on ingenuity of urban Africa. If we bring these negotiations over materials and environments back into the story of the ongoing struggle for decolonization, then engagements with failure are anything but an Afropessimist narrative. Instead, they demand we just as seriously engage with acts of provisioning as we do with the idealized futures of planners and experts that go ultimately unfulfilled. Again, to quote Magaziner, writing history requires that we see time as "not an inert medium through which trends and ideologies pass and are transmitted . . . time must be understood to be soil." In soil, you can plant—perhaps the quintessential way we humans plan for the future—but you cannot always know what might come up. Considering these moments of recalibration is vitally important not just for how we narrate the past but also for how we use the past to consider an increasingly unknown planetary future. Rather than predicting a future spent in a state of ongoing crisis, historians that highlight the ways communities have shaped the landscapes of African cities when plans go awry might instead suggest that urban Africans have a deep capacity to solve problems and reimagine the present.

NOTES

Introduction

1. Dar es Salaam means "haven of peace" in Arabic. For more on Dar's nineteenth-century history see Sutton, "Dar es Salaam," 1–18. See also Brennan, *Taifa*, 58–70. For a history of colonial urban planning, see Brennan, Burton, and Lawi, *Dar es Salaam*.

2. In Kiswahili, the abstract collective noun takes a *U* before it, so *Zaramo* becomes *Uzaramo*.

3. The Germans originally chose Bagamoyo.

4. Some other works of scholarship on urban-rural linkages include: Englund, "Village in the City, The City in the Village," 135–54; Mbilinyi, "'City' and 'Countryside' in Colonial Tanganyika," WS88–96; Ferguson, *Expectations of Modernity*; Geschiere and Gugler, "Urban-Rural Connection," 309–19; Hart, "Migration and Tribal Identity among the Frafras of Ghana," 21–36; De Boek, "Spectral Kinshasa," 309–28; and Mercer, "Landscapes of Extended Ruralisation," 72–83. More globally, there is also increasing attention being given to the growing reality of "urbanized countryside" or "ruralized cities." German urban theorist Thomas Sieverts uses the term *Zwischenstadt*, or "between city" to describe these environments in Davis, *Planet of Slums*, 9.

5. When not explicitly segregated by race, as was the case under the Germans, planning documents still entrenched race in neighborhoods by zoning for density. See Armstrong, "Colonial and Neocolonial Urban Planning," 43–66.

6. For example, see Molohan, *Detribalization*.

7. Brennan, *Taifa*; Ivaska, *Cultured States*; Callaci, "Street Textuality," 183–210.

8. Cronon, *Nature's Metropolis*; and Williams, *Country and the City*.

9. Leslie, *Survey of Dar es Salaam*, 93.

10. Frederick Cooper, writing about East Africa's economic and labor history, has called this "straddling." Cooper, "Africa and the World Economy" 1–86.

11. An early and foremost example is Raymond Williams's *The Country and the City*. See also a more recent collection on the topic: Kneitz, ed. "The Country and the City," special issue, *Global Environment* 9, no. 1 (2016).

12. I will be utilizing the term of the time, the "Third World," throughout. While clearly a problematic term that ordered and othered nations in the context of the Cold War, it was also taken up by members of the Global South as an invocation of political and economic solidarity. No other term adequately depicts the presumed positioning of nations like Tanzania in the postcolonial period while also chronicling a history of how these nations sought to upend a global order that had left them "behind" and made them dependent on the "First World's" economic imperatives.

13. Davis, *Planet of Slums*, 19. See also Sassen, "Cityness in the Urban Age," 1–3.

14. Robinson, "Global and World Cities," 545.

15. Robinson, "Global and World Cities," 531.

16. To borrow James Ferguson's term from *Expectations of Modernity* (1999).

17. While environmental historians have not readily take up the histories of cities in the Global South (a recent exception is Nagendra, *Nature in the City*, there are still many important and rich accounts of urban environments and urban infrastructures. See Bryant and Bailey, *Third World Political Ecology*; Martinez-Alier, "Ecology and the Poor," 621–39; and Heynen, Kaika, and Swyngedouw, *In the Nature of Cities*. In anthropology, a burgeoning scholarship on infrastructure, particularly in developing world cities, has been very insightful. See Nikhil Anand, *Hydraulic City*; Weinstein, *Durable Slum*; Herrera, *Water and Politics*; Fredericks, *Garbage Citizenship*; Larkin, *Signal and Noise*; von Schnitzler, *Democracy's Infrastructure*.

18. See, for example, Gandy, *Fabric of Space*; Heynen, Kaika, and Swyngedouw, *In the Nature of Cities*; and Davis, *Planet of Slums*. Simone, *For the City Yet to Come*.

19. See, for example, McNeill and Engelke, *Great Acceleration*; Kaplan, "Coming Anarchy." *Lagos/Koolhaas*, dir. Bregtje an der Haak.

20. In 1969, the distance from the edge of the city to its center ranged from 6 to 10 kilometers and rose to 15 kilometers in 1978. By the late 1990s it was around 30 kilometers. Diaz Olvera, Plat, and Pochet, "Transportation and Access to Urban Services in Dar Es Salaam." For more on Dar's spatial expansion see Mkalawa and Haixiao, "Dar es Salaam City Temporal Growth and Its Influence on Transportation." For a more recent overview of Dar's growth see Sam Sturgis, "The Bright Future of Dar es Salaam, an Unlikely African Megacity," *CityLab*, February 25, 2015, accessed August 16, 2019, https://www.citylab.com/design/2015/02/the-bright-future-of-dar-es-salaam-an-unlikely-african-megacity/385801/.

21. For statistics on Dar's population growth see Andreasen, "Population Growth and Spatial Expansion of Dar Es Salaam," 9. At the time of Dar growing at a 7.8 percent rate, national population growth rate was around 3 percent.

22. Important books that have helped shape this new vision of African urbanism include Simone, *City Life from Jakarta to Dakar*; Simone and Pieterse, *New Urban Worlds*; Simone, *For the City Yet to Come*; Myers, *African Cities*; Parnell and Pieterse, *Africa's Urban Revolution*. While not about African cities, another important collection is Roy and Ong, *Worlding Cities*.

23. For a good essay on this new scholarship see the introduction to Diof and Fredericks, *Arts of Citizenship in African Cities*.

24. These titles include De Boeck, *Kinshasa*; and Nuttall and Mbembe, *Johannesburg*.

25. George, *Making Modern Girls*; Jean-Baptiste, *Conjugal Rights*; Ivaska, *Cultured States*; Callaci, *Street Archives and City Life*; Brennan, *Taifa*; and Fair, *Reel Pleasures*.

26. There has been, however, quite a good collection of work in the past thirty years on Dar es Salaam written by planning scholars. For example, see the works of W. J. Kombe, J. M. Lusugga Kironde, Wolfgang Scholz, and J. L. P. Lugalla. There are also several anthropologists also doing important work on urban environments as noted in an earlier footnote (17).

27. Rademacher, "Urban Political Ecology," 139.

28. McNeill and Engelke, *Great Acceleration*.

29. Roy and Ong, *Worlding Cities*. Roy and Ong get "specter of comparison" from Anderson, *Spectre of Comparison*.

30. For some of the literature that highlight urban crisis in African cities: Stren and White, *African Cities in Crisis*; Tostensen, Tvedten, Vaa, *Associational Life in African Cities*; Konings and Foeken, *Crisis and Creativity*. Enwezor et al., *Under Siege, Four African Cities*.

31. Jackson, "Rethinking Repair," 221–40.

32. For a good summary of this scholarship see Lockrem and Lugo, "Infrastructure." See also recent edited volumes including Collier, Mizes, and von Schnitzler, eds., "Public Infrastructures and Infrastructural Publics"; and Anand, Gupta, and Appel, *Promise of Infrastructure*.

33. For a critique of the notion of resilience, a term frequently used to describe urban Africans, see Evansand Reid, *Resilient Life*.

34. Serlin, "Confronting African Histories of Technology," 97.

35. White, "Hodgepodge Historiography," 314–15.

36. Simone, *For the City Yet to Come*, 1.

37. As I will explain in the next chapter, for much of the 1970s the Municipal Council of Dar was dissolved by the state. The Tanzanian National Archives does not have records (to my knowledge) from ministries such as the Ministry for Lands, Housing and Urban Development and the ministries themselves no longer seem to have those documents from the 1970s and 1980s either.

38. For these twenty-six interviews conducted in October–December of 2009, I designed my interview questions akin to what Jamie Monson and Susan Geiger have called "modified life histories." See Monson, "*Maisha*," 315. Geiger, *TANU Women*.

39. Many Tanzanian historians in recent years have turned to newspapers as a vibrant source for cultural history. Andrew Ivaska, in his book *Cultured States*, argues that using newspapers and other media crucially destabilizes the entrenched authority of oral sources within African history while offering another way to recover Af-

rican perspectives. For more on using newspapers in Tanzanian history see Hunter, *Political Thought and the Public Sphere in Tanzania*, as well as Peterson, Hunter, and Newell, *African Print Cultures*.

40. Ivaska, *Cultural States*, 31. Also, according to Martin Sturmer, in a 1968 survey, 44 percent of urbanites in Dar claimed to read the paper daily. This statistic does not also include the frequency with which newspapers may have also been read aloud to small groups of people across the city. It is important to realize how avid Tanzanians were for news and rumors. Sturmer, *Media History of Tanzania*, 110.

41. Konde, *Press Freedom in Tanzania*, 135.

42. For more on these sorts of environmental interventions see Gissen, *Subnature*.

43. This archive I assembled through visits to the University of Dar es Salaam; United States National Archives, College Park, Maryland; International Monetary Fund Archives, Washington DC; World Bank online archives; British National Archives, Kew Gardens; British Library for Development Studies, University of Sussex, UK.

Chapter 1: Decentering Dar

1. From 1957 to 1967, Dar grew from 93,363 inhabitants to 272,821. Brennan, Burton, and Lawi, *Dar es Salaam*, 53.

2. Mabogunje, "Urban Planning and the Post-colonial State in Africa," 140.

3. Segal, "Ethnic Variables in East African Urban Migration," 194–204. Singer, "Rural Unemployment as a Background to Rural-Urban Migration in Africa," 37–45. Byerlee, "Rural-Urban Migration in Africa," 543–66. Potts, "Shall We Go Home?," 245–64. Bates, *Markets and States in Tropical Africa*.

4. As the end of the studio era arrived in Hollywood, the bleak landscapes of American and European cities became the new tableau for a very different version of the Hollywood movie in the 1970s. Webb, *Cinema of Urban Crisis*.

5. There is much written on the New Town movement. For a recent example, see Wakeman, *Practicing Utopia*.

6. Sabin, *Bet*, 1.

7. For more on Ehrlich, see also Robertson, *Malthusian Moment*; and Robertson, "This Is the American Earth," 561–84.

8. Ehrlich, *Population Bomb*, 1.

9. Schumaker, *Small Is Beautiful*. Immerwahr, *Thinking Small*.

10. Mazrui, "Tanzaphilia," 20–26.

11. In 1977 it was renamed Chama cha Mapinduzi (CCM) when it merged with the Afro-Shirazi party. Tanzania at this time was a one-party state.

12. Nyerere, *Arusha Declaration*.

13. Tanzania was a popular aid recipient despite the announcement of state socialism, in part because the nation remained nonaligned during the Cold War. For more on *ujamaa*, see Lal, *African Socialism in Postcolonial Tanzania*; Scott, *Seeing Like a State*; Schneider, *Government of Development*.

14. *Nationalist*, September 24, 1969, as quoted in Schneider, *Government of Development*, 95.

15. Smith, "We Can't Go to the Moon!," 42–43.

16. "Mwalimu Deplores Laziness," *Daily News*, August 23, 1972.

17. Nyerere, *Ujamaa Vijijini*.

18. Stren, "Ujamaa Vijijini and Bureaucracy in Tanzania," 593.

19. Schneider, *Government of Development*, 89. For more on the number of Tanzanians relocated (which has been widely debated) see Yeager, "Demography and Development Policy in Tanzania," 500.

20. Brennan, *Taifa*, 173.

21. Lal, *African Socialism in Postcolonial Tanzania*, 30.

22. John Lonsdale, "Moral Economy of Mau Mau," 265–504; Peterson, *Ethnic Patriotism and the East African Revival*; White, *Comforts of Home*.

23. Brennan, "Blood Enemies," 293.

24. Maddox, "Leave Wagogo, You Have No Food."

25. Mesaki and Nimtz, "Politics of Capital Relocation," 3.

26. J. Seng'enge, "Defer This Project to the Right Time," *Daily News*, September 20, 1972. Cyril Gabby Mgaya, "Let's First Do First Things," *Daily News*, September 22, 1972.

27. Christopher Mtapanta, "We Need a Modern Capital," *Daily News*, October 10, 1972.

28. Mesaki and Nimtz, "Politics of Capital Relocation," 20.

29. Mabogunje, "Urban Planning and the Post-Colonial State in Africa," 140.

30. Ngaranaro, "Dar es Salaam is Ageing," *Daily News*, October 30, 1972.

31. Mesaki and Nimtz, "Politics of Capital Relocation," 8.

32. While voted on in 1972, the idea was brought up six years prior by Nyerere. Mesaki and Nimtz, "Politics of Capital Relocation," 2.

33. *Why Dodoma?* pamphlet, Capital Development Authority, 1973.

34. Simon Peter Mlenge, *Daily News*, September 18, 1972.

35. Mesaki and Nimtz, "Politics of Capital Relocation," 10.

36. Mesaki and Nimtz, "Politics of Capital Relocation," 26.

37. See for example Sugrue. *Origins of the Urban Crisis*. Harvey, *Social Justice in the City*.

38. For an example of this medical language referring to urban crisis see Gruen, *Heart of Our Cities*.

39. It was originally supposed to be seven villages but ultimately ended up with five.

40. Julian Spector, "Why Reston, Virginia, Still Inspires Planners 50 Years Later," *CityLab*, March 23, 2016, http://www.citylab.com/design/2016/03/reston-virginia-urban-planning-suburbs-robert-simon/474729/,

41. As Sophie van Ginneken points out, it is likely that Dodoma is also modeled after Don Mills, a suburb of Toronto also built in the New Town model and planned

by Macklin Hancock. "The Burden of Being Planned. How African Cities Can Learn from Experiments of the Past: New Town Dodoma, Tanzania," International New Town Institute, accessed September 9, 2017, http://www.newtowninstitute.org/spip.php?article1050.

42. Michael T. Kaufman, "Tanzania's Official Capital, Dodoma, Is Still Just a Magnificent Dream," *New York Times*, January 18, 1977, 2.

43. Friedman, "Global Postcolonial Moment and the American New Town," 553–76.

44. Hull, "Communities of Place, Not Kind," 769.

45. For more on postwar village development, see Immerwahr, *Thinking Small*; Callaci, "'Chief Village in a Nation of Villages,'" 96–116; and the work of Otto Koenigsburger and Albert Mayer in India, such as Koenigsberger, "New Towns in India," 94–132; and Mayer, *Pilot Project, India*. For more on boundary objects, see Star and Griesemer, "Institutional Ecology, 'Translations' and Boundary Objects," 393.

46. For a good discussion of Dodoma's plan see Emily Callaci's article "'Chief Village in a Nation of Villages.'"

47. Kaufman, "Tanzania's Official Capital," 2.

48. For more on villagization in Dodoma region see Schneider, *Government of Development*.

49. Capital Development Authority, *Why Dodoma?*

50. Project Planning Associates, *National Capital Master Plan, Dodoma, Tanzania*.

51. Press release, "Mwalimu Approves Interim Plan for Dodoma," FCO 31/1757: Change of Capital City in Tanzania," National Archives, Kew Gardens, UK.

52. Vale, *Architecture, Power, and National Identity*, 151.

53. Hoyle, "African Socialism and Urban Development," 213–14.

54. Lupala and Lupala, "Conflict between Attempts to Green Arid Cities and Urban Livelihoods," 26.

55. This policy lasted for the ten years it took to finish the capital. "Dodoma Housing Problem," *Daily News*, August 6, 1976. Kanyama, "Can the Urban Housing Problem be Solved Through Physical Planning?," 59.

56. Kulaba, *Housing, Socialism and National Development in Tanzania*, 55.

57. Kulaba, *Housing, Socialism and National Development in Tanzania*, 67.

58. John M. Sankey, "Sankey, John Anthony CMG (b 1930)," transcript of oral history interview, London, April 19, 2010, Lancaster Gate, British Diplomatic Oral History Programme, Churchill College, Cambridge, accessed August 2018, https://www.chu.cam.ac.uk/media/uploads/files/Sankey.pdf.

59. McAuslan, "Law and Lawyers in Urban Development," 64.

60. McAuslan, *Ideologies of Planning Law*, xi–xii.

61. Holston, *Modernist City*, 5.

62. Nyerere, *Decentralization*, 1.

63. "Every Tanganyika citizen is an integral part of the nation and has the right to take an equal part in government at local, regional and national level." 'The National

Ethic,' in "Report of the Presidential Commission on the Establishment of a Democratic One Party State," reprinted in Cliffe, *One-Party Democracy*, 440.

64. See Samoff, "Bureaucracy and the Bourgeoisie," 30–62.

65. It would not be until 1978, under worsening urban conditions, that the council would be reinstated.

66. Kironde, "Evolution of Land Use Structure of Dar es Salaam," 92.

67. Hyden, *Beyond Ujamaa in Tanzania*, 135.

68. For more on this, see the subsequent chapters in this book, but also Mesaki, "Operation Pwani," and Owens, "From Collective Villages to Private Ownership."

69. But by 1952, wage labor still made up less than 10 percent of the economically active population across Tanzania. Over 50 percent of those jobs were in agriculture. If and when people migrated during the colonial period it had predominantly been to other rural areas for agricultural labor. Sabot, *Economic Development and Urban Migration*, 21. See also the five-part study Bienefeld and Sabot, "National Urban Mobility Employment and Income Survey of Tanzania." Some parts are available at Ardhi University Library, University of Dar es Salaam.

70. Sabot, *Economic Development and Urban Migration*, 45.

71. Bienefeld and Sabot, "National Urban Mobility Employment and Income Survey of Tanzania," part 1.

72. Sabot, *Economic Development and Urban Migration*, 69.

73. Mabogunje, "Urban Planning and the Post-Colonial State in Africa," 153.

74. Cronon, *Nature's Metropolis*.

Chapter 2: Belongings

Epigraph: Benjamin, *Reflections*, 135–36.

1. The notion of sedimentation comes from Ann Laura Stoler's introduction to *Imperial Debris: On Ruins and Ruination*, where she notes that the "uneven temporal sedimentations" of empire give "contour and carve through the psychic and material space in which people live and what compounded layers of imperial debris do to them" (2).

2. He continues, "Not to belong is to be constantly vulnerable to the accusation of trespass—even when in legalistic terms it is utterly groundless." Crowley, "Politics of Belonging," 17.

3. This chapter discusses only material and spatial belonging in the city, not cultural, social, or religious belonging. For other forms of cultural and political "belonging" and how they have shaped Tanzanian history see Lal, "Militants, Mothers, and the National Family," 1–20; Hunter, *Political Thought and the Public Sphere in Tanzania*; Ivaska, *Cultured States*; Askew, *Performing the Nation*; Fair, "Drive-In Socialism," 1077–104; Brennan, *Taifa*.

4. See Geschiere and Gugler, "Urban-Rural Connection," 309–19. Ferguson, "City and Country in the Copperbelt," 80–92; Mbilinyi, "'City' and 'Countryside' in Colonial Tanganyika," WS88–96; Southall and Gutkind, *Townsmen in the Making*; Powdermaker, *Copper Town*; Mitchell, *Kalela Dance*.

5. These forms of belonging are sometimes referred to as *elective belonging*, which entails cultivating a "sense of spatial attachment, social position, and forms of connectivity to other places." Savage, Bagnall, and Longhurst, *Globalization and Belonging*, 29.

6. Brennan, *Taifa*, 19. Brennan defines urban entitlement as "macroeconomically speaking, the rural subsidization of urban life."

7. Brennan, *Taifa*, 85. Even further back, many of these laws have their roots in ideals of British town planning. Scholz, Robinson, and Dayaram, "Colonial Planning Concept and Post-colonial Realities," 67–94.

8. Dar for much of the colonial period remained quite small, with only 20,000 residents in 1921. But that population had more than doubled by 1948, to 50,765.

9. Shivji et al., *Report of the Presidential Commission*, 65.

10. This is not unique to Tanzania. See for example, Cooper, *On the African Waterfront*.

11. Leslie, *Survey of Dar es Salaam*.

12. Leslie, *Survey of Dar es Salaam*.

13. "Notes Leading up to Recommendations that a Sociological survey of Dar es Salaam should take place," "Dar es Salaam Extra Provincial District" District Book, n.d. (but from the 1940s or 1950s), accession #550 Home Affairs HQ Dar Microfilm 30, Tanzania National Archive, Dar es Salaam.

14. Brennan, *Taifa*, 88. Frederick Cooper, who has written about "the labor question" in the late colonial city, has noted that the colonial state transitioned to permanent labor rather than day labor and encouraged African men to bring their families with them into cities as an attempt to alleviate fears of labor unrest. Brennan notes that for Dar es Salaam it was more complicated than just permanent labor because the city was so expensive. Cooper, *Decolonization and African Society*.

15. Brennan, *Taifa*, 93.

16. Ilala would be developed through government housing projects while Temeke would be private construction. Magomeni, in contrast, was to be a mix including its existing population as well as new middle-class housing.

17. "Letter from the Director of Town Planning to the Chief Architect," October 23 (year illegible), accession no. 359 (Secretariat File # 32575/2), African Housing Schemes Magomeni File, Tanzania National Archives (TNA).

18. Brennan, *Taifa*, 91. Also Mbilinyi, "'City' and 'Countryside' in Colonial Tanganyika," WS93.

19. Kironde, "Race, Class and Housing in Dar es Salaam," 110.

20. D.D., "African Housing Schemes Magomeni File Notes" May 17, 1952, accession no. 359 (Secretariat File 32575/2), TNA. For more on the poor infrastructure of "native" areas of town during the colonial period, see Burton, *African Underclass*, 88–91.

21. Brennan, *Taifa*, 42.

22. Brennan, *Taifa*, 42.

23. See Burton, *African Underclass.*

24. Simone, "On the Worlding of African Cities," 21.

25. Provincial Commissioner to District Commissioner of Kisarawe, "Applications for RO in the chiefdom of Kunduchi, July 1950," July 12, 1950, accession no. 11/502, TNA.

26. Provincial Commissioner to District Commissioner of Kisarawe, "Applications for RO in the chiefdom of Kunduchi, July 1950," July 12, 1950, accession no. 11/502, TNA.

27. Shivji et al., *Report of the Presidential Commission,* 71.

28. Shivji et al., *Report of the Presidential Commission,* 68. The act also legislated the city's rules for land use, density of occupancy, guidelines for construction materials, physical infrastructure of roads, and preservation of open space, and remained unaltered until 2007. Scholz, Robinson, and Dayaram, "Colonial Planning Concept and Post-Colonial Realities," 67–94.

29. Shivji et al., 66.

30. Swanson, "Sanitation," 387–410.

31. Municipal Council of Dar es Salaam, Reports to Committees September 1957 Dar es Salaam City Council Minutes Appendix 83, 275, Tanzania National Library, Dar es Salaam.

32. Project Planning Associates, "Recommended Capital Works Programme," 5.

33. PADCO, *Proposal for an Urban Development Corporation in Tanzania,* 20.

34. Known officially as the 1971 Acquisition of Buildings Act. See James Brennan and Andrew Burton, "The Emerging Metropolis: A History of Dar es Salaam, circa 1862–2000," in Brennan, Burton, and Lawi, *Dar Es Salaam,* 56.

35. Aminzade, *Race, Nation, and Citizenship in Post-colonial Africa,* 226.

36. For more on this see Vassanji, *And Home Was Kariakoo,* 7. See also Aminzade, *Race, Nation, and Citizenship,* 220–30, for a discussion of the nationalization of commercial and rental properties.

37. Tarimo, *Effects of Trade Liberalisation on Property Development,* 86.

38. There will be more on this in the next chapter, but the Swahili house of Dar was defined as a house where the "walls may be built of wood and earth or of stones and plaster, roofs of coconut fronds (makuti), opened up petrol tins (madebe) or corrugated iron (mabati). Sutton, "Dar es Salaam," 14.

39. World Bank, *Tanzania Appraisal of National Sites and Services Project,* 7.

40. World Bank, *Tanzania Appraisal of National Sites and Services Project,* 7. Within development circles this policy was considered at the time quite progressive at the time and was executed with help of the World Bank who drew up plans for two different Sites and Services Upgrading programs through the 1970s.

41. For more on this see Stren, "Underdevelopment, Urban Squatting, and State Bureaucracy," 81.

42. Nnkya, *Shelter Co-operatives in Tanzania,* 45.

43. Mwakysusa, "Penetration of TANU into Industries," 4.

44. Coulson, *Tanzania*, 329.

45. Mihyo, *Women, Work and Struggle*, 38. Of the 153 sent abroad, none were women.

46. Swantz and Bryceson, *Women Workers in Dar Es Salaam*, 8.

47. Alex Perullo, "Live from Dar es Salaam" in Brennan, Burton, and Lawi, *Dar es Salaam*. See also Perullo, *Live from Dar es Salaam*, 16.

48. NARA, USAID Lot No. 80-116, Box no. 22, Accession no 286-80-112, Folder SOC 5-Labor Survey of Tanganyika Textile Industries LTD, October 1969, National Archives, Maryland, USA.

49. While there were only three cooperatives in the beginning of the 1970s, by 1987 there were thirty-three. Huba, "Access to Housing through Cooperatives," 100–14.

50. J. M. Kerenge, "Housing Problems in Urban Centres in Tanzania," Rotterdam, Netherlands, Bouwcentrum International Education Rotterdam, Working Paper Series No. 573, 4, available in the collection of Ardhi University Library, Dar es Salaam, Tanzania. This initial collaboration was also funded in part by the World Bank.

51. Nnkya, *Shelter Co-operatives in Tanzania*, 68.

52. Ndatulu and Makileo, *Housing Cooperatives in Tanzania*, 32.

53. Brennan, *Taifa*, 188.

54. Richard Stren, "Underdevelopment, Urban Squatting and State Bureaucracy," 75. Stren gets this figure from the Ministry of Lands, *Housing and Urban Development Manzese Social Survey* (Dar es Salaam: Government of Tanzania, 1975).

55. Stren, "Underdevelopment, Urban Squatting and State Bureaucracy," 80.

56. Stren, "Underdevelopment, Urban Squatting and State Bureaucracy," 76.

57. World Bank, *Report and Recommendation of the President for the Reconstitution and Development to the Executive Directors on a Proposed Loan to the United Republic of Tanzania for an Urban Water Supply Project*," December 7, 1976, Report No. P-1928a-TA, http://documents.worldbank.org/curated/en/469301468116372821/pdf/multi0page.pdf

58. Stren, "Underdevelopment, Urban Squatting and State Bureaucracy," 78.

59. In practice, however it was not always groups of ten that made up these units. O'Barr, "Cell Leaders in Tanzania," 438.

60. O'Barr, "Cell Leaders in Tanzania," 438. It should be noted that while their particular use may have been uniquely developed during Ujamaa, ten cells were not unique to Tanzania. Parker Shipton notes their use in Kenya in *Mortgaging the Ancestors: Ideologies of Attachment in Africa* (2009), 104. Nelson Mandela in his autobiography notes their use in urban townships in South Africa by the ANC and more generally the "neighborhood unit" was a concept utilized in American planning in the 1920s by Clarence Perry and was spread to Indian planning via the German planner Otto Koenigsberger. Koenigsberger noted that it "had special appeal to people

of undeveloped countries" because they "form the best possible links with the type of community life they know from their villages." See Hull, "Communities of Place, Not Kind," 767.

61. *Wabalozi* is the plural of *mbalozi*. Hull, "Communities of Place, Not Kind," 455. For more on ten cells see also Jennings, "'Development is Very Political in Tanzania,'" 76–92; and Levine, 'TANU Ten-House Cell System."

62. Health Week Sub-Committee, Minutes of the Meeting Held in the committee room, Karimjee Hall, 12th August, 1970, at 4:30pm, City Council of Dar es Salaam, Report of Committees, printed by Council's Rotaprint Town Clerk's Office, DSM, Tanzania National Library.

63. In O'Barr's research in Usangani there was a long history of political activism that made women some of the hardest-working and effective ten cell leaders. "Female informants often express the belief that they try hard, harder than men, to see that cells become effective networks." O'Barr, "Cell Leaders in Tanzania," 460–61.

64. Health Week Sub-Committee, Minutes of the Meeting Held in the committee room, Karimjee Hall, 12th August, 1970, at 4:30pm, City Council of Dar es Salaam, Report of Committees, printed by Council's Rotaprint Town Clerk's Office, DSM, Tanzania National Library.

65. Agrawal, *Environmentality*.

66. Kombe and Kriebich, "Informal Land Management in Tanzania and the Misconception about its Illegality," 3.

67. Mohammed Halfani, interview with the author, Makongo Juu, Dar es Salaam, June 13, 2014.

68. Mohammed Halfani, interview with the author.

69. "Idlers are Enemies," *Daily News*, July 4, 1976.

70. Kombe, "Land Use Dynamics in Peri-urban Areas," 118.

71. Owens, "From Collective Villages to Private Ownership," 219.

72. "Full Scale War on Loiterers Planned," *Daily News*, November 12, 1976.

73. Burton, "Haven of Peace Purged," 137.

74. In contrast, Rwegasira claimed that married women, children, and disabled people would remain unharassed. "Dar Jobless Campaign," *Daily News*, November 29, 1976.

75. For more on the politics and anxieties of women in Dar see Ivaska, *Cultured States*, 86–123.

76. "Dar Drive to Resettle Jobless Begins," *Daily News*, November 17, 1976.

77. "Dar Region Resettles over 2500 Families," *Daily News*, November 30, 1976.

78. See "Prostitutes Have Identity Cards!" letter by E Makundi in *Daily News*, November 18, 1976.

79. Felix Kaiza, "Getting a House in Dar es Salaam," *Daily News*, June 1972.

80. James Mpinga, "How Dar Regions Resettled the Unemployed," *Daily News*, November 24, 1976.

81. Mpinga, "How Dar Regions Resettled the Unemployed."

82. Charles Rajabu, "Dar Jobless Campaign: Over 3000 Families Resettled," *Daily News*, December 11, 1976.

83. James Mpinga, "Jobless 'Will Be Made to Work,'" *Daily News*, November 20, 1976.

84. Tony Barros, "Fimbo ya Monyonge: The Film of Our Time Is Here," *Daily News*, February 23, 1976. The book version was printed two years later in 1978 and there was an accompanying *Daily News* article for its publication that spared no details and summarized the entire plot in the newspaper. "Poor Man's Salvation," review by Isaac Mruma, September 2, 1978.

85. Fair, *Reel Pleasures*, 190.

86. See Bertoncini-Zúbková, *Outline of Swahili Literature* for many examples, including Godfrey Nyasula, *Laana ya Pandu* (1974), Alex Banzi's *Zike Mwenyewe* (1977), Ndyanao Balisidya, *Shida* (1975), Jerome Ngalimecha Ngahyoma's play *Huka* (1973), and Emmanuel Mbogo, *Giza Limeingia* (1980). To be clear, there were also other genres of fiction, particularly detective fiction that celebrated urban life and developed a particularly Tanzanian aesthetic of belonging in the city. For more on this, see Emily Callaci's recently published book, *Street Archives and City Life*.

87. "Fimbo ya Mnyonge at Ilala December 5 1976," *Daily News*, December 3, 1976.

88. So brief is the ride in fact that the 1968 Master Plan had plans for making a bridge to span the distance. A bridge was finally finished in 2016.

89. "Mziki wa dansi zilipendwa," https://www.youtube.com/watch?v=7ig5SdJnua8.

90. Kijakazi Kyelula, "Geza Ulola is Dar's best Ujamaa Village," *Daily News*, July 16, 1976.

91. Kyelula, "Geza Ulola is Dar's best Ujamaa Village."

92. "Dar Pledges War on Jobless," *Daily News*, November 18, 1976.

93. Mpinga, "How Dar Regions Resettled the Unemployed."

94. Rajabu, "Dar Jobless Campaign."

95. "Full Scale War on Loiterers Planned," *Daily News*, November 12, 1976,

96. Mutasingwa, "History of Dar es Salaam," 41.

97. Mutasingwa, "History of Dar es Salaam," 41.

98. "Five Hundred Families Resettled in Dar Region," *Daily News*, November 22, 1976.

99. Ally Kinongo, interview with the author, Mbagla, 2009.

100. For more on why people moved to the suburbs and how it was not a sign of "poor or incomplete adaptation to urban life" see Owens, "Post-colonial Migration," 249–74.

101. Hamada Ali Mnora, interview with the author, Mbagala, October 30, 2009.

102. Kombe, "Land Use Dynamics in Peri-urban Areas," 126.

103. TNA File No 1/26 Wakiliate Kunduchi-Development: "Closing of are to alienation on new DSM/Morogoro Road 25 Feb 1954.

104. Owens, "From Collective Villages to Private Ownership," 217.

105. "Projects in the Pipeline," *Daily News*, April 30, 1976.

106. "Projects in the Pipeline," *Daily News*, April 30, 1976.

107. Issa Shivji et al, *Report of the Presidential Commission*, 62–63. However, this might actually have been 1979 when the new master plan designated Mbezi a planning area. See Kombe, "Demise of Public Urban Land Management and the Emergence of Informal Land Markets in Tanzania," 27.

108. Brennan, *Taifa*, 42.

109. Mushi, "Regional Development Through Rural Urban Linkages," 48.

110. McNeur, *Taming Manhattan*, 3.

Chapter 3: Building

1. This seems to be a play on "trouble at t' mill," a phrase associated with labor unrest in northern England.

2. Saul, *Flawed Freedom*, 34.

3. J. K. Nyerere, speech to the World University Service, June 27, 1966, reproduced in the *Standard*, June 28, 1966.

4. T. I. Saver, "Better Burnt Bricks," *Daily News*, January 30, 1975.

5. Stoler, *Imperial Debris*.

6. Larkin, *Signal and Noise*, 28.

7. Low and Lonsdale, "Towards the New Order." As Lynn Thomas notes (citing Frederick Cooper), "British and French colonial officials used 'modern' in the immediate post-World War II period to describe more interventionist policies." Thomas, "Modernity's Failings, Political Claims, and Intermediate Concepts," 734.

See for example Burke, *Lifebuoy Men, Lux Women*; Martin, "Contesting Clothes in Colonial Brazzaville," 401–26. Thomas, "Modern Girl and Racial Respectability in 1930s South Africa," 461–90.

8. Here I am following Larkin, who draws on Patrick Joyce and Timothy Mitchell's definition of technopolitics. "For Joyce techno politics rests on the idea that liberalism is a mode of politics that function through invisibility—meaning the lack of overt intervention by governmental bodies in everyday affairs. In their place liberalism seeks proxies in technological regimes—building sewers, organizing libraries, mapping, counting censuses—which are political precisely because they are seen as technical and so outside of political processes." Larkin, *Signal and Noise*, 47.

9. *Wives of Nendi*, dir. Peet. See also Burke, *Lifebuoy Men, Lux Women*.

10. Kent, *Domestic Architecture and the Use of Space*, 122.

11. There was also, as mentioned in the previous chapter, some areas within these neighbourhoods set aside for "self-building" but only granted highly selectively by those who demonstrated they had stable incomes and families. It is also worth noting

that before 1945, the colonial administration did allow Africans to build in any style in the central African neighborhoods of Ilala and Kariakoo.

12. Leslie, *Survey of Dar es Salaam*, 153–54.

13. This prohibition was not always enforced.

14. Vestbro notes that the house style that has come to be commonly called a "Swahili-style house" was likely first a Zaramo house style. Vestbro, *Social Life and Dwelling Space*.

15. Section titled "What is a House" in Extra Provincial District Book, TNA Archives, n.p.

16. "What is a House" notes that one house of twenty had "seven tribes in six rooms."

17. "What is a House."

18. For more on this, see Morton, *Age of Concrete*.

19. Hanlon, "Where Concrete Meets Cane," 627–29. Angola's colonial capital, Luanda, was also demarcated materially by sand and asphalt. Moorman, *Intonations*.

20. Hanlon, "Where Concrete Meets Cane," 627–29.

21. It is important to note here that there were African neighborhoods in the city limits as well.

22. Chang, "Building a Colonial Technoscientific Network," 217.

23. Chang, "Building a Colonial Technoscientific Network," 223.

24. File No CO 822/589 "African Urban Housing in Tanganyika" British National Archives, Kew Gardens.

25. April 1, 1953 Anthony Atkinson to "Rogers" regarding Atkinson's trip to East Africa and his assessment of African housing needs. File No CO 822/589 "African Urban Housing in Tanganyika," British National Archives, Kew Gardens.

26. Gutschow, "*Das Neue Afrika*," 257.

27. Gutschow, "*Das Neue Afrika*," 256–258.

28. Cloete, "End of Era with Threat of Jungle Taking Over."

29. Stoler, *Imperial Debris*. For a different example of this see Stephen J. Collier's notion of "the intransigence of things" in his book *Post-Soviet Social: Neoliberalism, Social Modernity, Biopolitics* (2011).

30. "Imperial formations are defined by racialized relations of allocations and appropriations. Unlike empires they are processes of becoming, not fixed things." Stoler, *Imperial Debris*, 8.

31. Julius Nyerere, 1977, quoted in Reginald Herbold Green, "Africa and the 1980s: Issues, Problems and Prospects in Rweyemamu's African Development Strategies in the Eighties," manuscript, Reginald H. Green Archive, OpenDocs, https://opendocs.ids.ac.uk/opendocs/handle/20.500.12413/5923?show=full.

32. Burchell, Gordon, and Miller, *Foucault Effect*.

33. Hutton, *Urban Challenge in East Africa*, 167. See also Swanson, "Sanitation Syndrome," 387–410.

34. Sunseri. *Wielding the Ax*, 149.

35. Sunseri. *Wielding the Ax*, 149.

36. Nyerere, *Freedom and Development*, 137.

37. To put this in context, one of the earliest campaigns of ujamaa was also to change the dress of rural Maasai see Schneider, "Maasai's New Clothes," 101–31.

38. Circular to All Regional Community Development Officers "Unicef Assistant to Self Help Schemes" from Community Development Division, 4th May 1964.

39. Freyhold, *Ujamaa Villages in Tanzania*, 132.

40. Freyhold, *Ujamaa Villages in Tanzania*, 132.

41. Daley, "Land and Social Change in a Tanzanian Village 1," 383.

42. Freyhold, *Ujamaa Villages in Tanzania*, 127.

43. Forty, *Concrete and Culture*, 15.

44. The former two by Amuli, the latter two by Almeida. For more on Almeida see Peter Burssens, "(Non)Political Position of the Architecture of Anthony B. Almeida."

45. *Many Words for Modern*, dir. den Hollander.

46. *Many Words for Modern*.

47. There is a recent and growing literature on cement focusing on its materiality and its politics. To name a few: Barry, *Material Politics*. See Forty, *Concrete and Culture*; Abourahme, "Assembling and Spilling-Over"; Harvey and Knox, "Abstraction, Materiality and the Science of the Concrete in Engineering Practice," 124–41. Morton, "Chamanculo in Reeds, Wood, Zinc and Concrete," 42–46.

48. Kuuya, *Import Substitution as an Industrial Strategy*, 24.

49. Kuuya, *Import Substitution as an Industrial Strategy*, 24.

50. Stakeholders: Cementia Holding, Zurich (40%), The Portland Cement Manufacturers LTD 40%, Smith Mackenzie 10%; the National Development Corporation (TZ gov) (10%) (Kuuya)

51. Vazifdar, "Cement Industry in Tanzania," 127.

52. In a 1973 study, the percentage was estimated to be as high as 40 to 50 percent of the production costs. "Using Burnt Bricks," *Daily News,* January 21, 1975.

53. To be exact, 44 percent. See Mwandosya and Luhanga, "Energy Use Patterns in Tanzania," 237.

54. Vazifdar, "Cement Industry in Tanzania," 130. The price went from 99 shillings per 1,000 liters to 529 shillings per 1,00 liters from 1974 to 1977.

55. Springer, "Wazo Hill Cement Works Construction."

56. Springer, "Wazo Hill Cement Works Construction."

57. Other words for cement are *sementi* or *saruji*.

58. "Mwalimu Opens Hydro-Electric Plant," *News Review,* January 1965, as cited in Hoag, *Developing the Rivers of East and West Africa*.

59. "Police Probe Dar Cement Plant Mess," *Daily News*, September 22, 1972.

60. "New Sabotage Claim at Cement Factory," *Daily News*, October 21, 1972.

61. Kuuya, *Import Substitution*, 24; and "Cement Output to Increase," *Daily News*, January 24, 1978.

62. "Cement Output to Increase," *Daily News*, January 24, 1978.

63. Yhdego, "Industrialization and Environmental Pollution in Tanzania," 78.

64. Eriken and Ronnstedt, "Building Cost Indices."

65. There were attempts to make more "vernacular" local cement in different regions.

66. For more on rumor and gossip in Africa see White, *Speaking with Vampires*.

67. Appropriate technology was a movement (some trace it back to Gandhi) that took off in the 1960s and 1970s after the publication of E. F. Shchumacher's book *Small is Beautiful*, in which he argues that in development contexts it is often more relevant and effective to find small, mobile, and cheap solutions to technological problems than solutions that are capital-intensive and frequently also require foreign expertise to implement and maintain. "Appropriate technology" took off in particular after the oil crisis, as will be discussed in chapter 6.

68. Nyerere, "Arusha Declaration Ten Years After," 29–31.

69. Wells, "Construction Industry in the Context of Development," 12.

70. Wells, "Construction Industry in the Context of Development," 14.

71. For more on this see Coulson, *Tanzania*, 343–46.

72. Wells, "Construction Industry in the Context of Development," 16.

73. Kaitilla, "Influence of Environmental Variables on Building Material Choice," 208.

74. For example, they also created reinforced sisal for roofing.

75. "Clampdown on Use of Cement and Iron Sheets," *Daily News*, December 28, 1973.

76. "Clampdown on Use of Cement and Iron Sheets," *Daily News*, December 28, 1973.

77. It should be noted that this is at this same time that housing out of "local materials" is being torn down in Dodoma. See Kulaba, *Housing, Socialism and National Development in Tanzania*, 60.

78. W. D. S. Mbaga, "Using Burnt Bricks," *Daily News*, January 21, 1975.

79. T. I. Svare, "Better Burnt Bricks," *Daily News*, January 30, 1975.

80. The factory was forecasted to make three hundred tons of clay products a day including bricks, tiles and building blocks. Capital Development Authority, *Building the National Capital*.

81. Humphrey, "Ideology in Infrastructure," 40.

82. Schwenkel, "Post/Socialist Affect," 254.

83. Svare, "Better Burnt Bricks."

84. Lawrence Kilimwiko, "Opting for Burnt Bricks," *Daily News*, July 26, 1978.

85. Kilimwiko, "Opting for Burnt Bricks."

86. Kaplinsky, *Economies of the Small*, 91. Chapter 4 of this book is a comparative look at brick manufacture in three African countries.

87. Pugu Brick Factory ran on oil from Dar's Tiper oil refinery.

88. Kaitilla, "Influence of Environmental Variables on Building Material Choice," 216.

89. "Using Burnt Bricks," *Daily News*, January 21, 1975.

90. "Using Burnt Bricks," *Daily News*, and Svare, "Better Burnt Bricks."

91. Mnzava, "Village Industries vs Savanna Forests."

92. Sunseri, *Wielding the Ax*, 158.

93. There is of course variation based on size and kind of tree but this is not noted.

94. Simon R. Nkonoki, *Energy Crisis of the Poor in Tanzania*, 33.

95. "Better Burnt Bricks," *Daily News*, January 30, 1975.

96. Immerwahr, *Thinking Small*, 74. Today, these industries in South Asia are sites of immense exploitation, particularly of child labor. See, for example, Humphrey Hawksley, "Why India's Brick Kiln Workers 'Live like Slaves,'" *BBC News*, January 2, 2014, sec. India, https://www.bbc.co.uk/news/world-asia-india-25556965.

97. For more on the contentious definition of voluntarism in nation building see Hunter, "Voluntarism, Virtuous Citizenship, and Nation-Building in Late Colonial and Early Postcolonial Tanzania," 43–61.

98. Nkonoki, *Energy Crisis of the Poor in Tanzania*, 32. See also Mnzava, "Village Industries vs. Savanna Forests."

99. Nkonoki, *Energy Crisis of the Poor in Tanzania*, 32.

100. While I have not been able to find substantial histories of brickmaking on Tanzanian missions it is clear that across much of Africa missions were built out of brick. The missions' use of bricks was not just about building in a "permanent" and recognizably Western material, but it was also pedagogical. It seems their very shape was also part of the aim to convert. John Mackenzie, writing about Nyasaland, notes that the Africans who worked on the construction of mission stations were known as "the bricks." He writes that "brickworks became the sine qua non of the mission station that demonstrated a full command of its environment. And being rectangular, it produced rectangular structures." This was in opposition to the round forms that Africans tended to build in. "One missionary in his memoir described the patience required to persuade an African bricklayer to lay his bricks in a straight line and to the vertical. Many commented on the sinuousness of African paths, and one traveler was even induced to say that elephants were better engineers than Africans." Porter, *Imperial Horizons of British Protestant Missions, 1880–1914*, 115.

101. Amiri S. Said, "NHC and Slum Clearing," *Daily News*, October 18, 1972.

102. Forty, *Concrete and Culture*, 52. But it is also thought of as something that "does not go to ruin" alluding to the longstanding view in French architectural circles that, as August Perret put it, "architecture is what makes beautiful ruins" (52).

103. Schmetzer, "Housing in Dar-es-Salaam," 506.

104. Nnkya, "Housing and Design in Tanzania," 31.

105. Dar es Salaam Ministry of Works, *Local Construction Industry Study: Summary Report*, 116. It is unclear if, at the time of this report this rule was still in place.

106. Kirobo, *Film-House in Manzese, Dar es Salaam*.

107. Hamada Ali Mnora, interview with the author, Mbagala, Dar es Salaam, Tanzania, October 2009.

108. Mzee Said Juma Akida, interview with the author, Mbagala, Dar es Salaam, Tanzania, October 2009.

109. Leslie, *Survey of Dar es Salaam*, 157.

110. World Bank, *Tanzania: The Second National Sites and Services Project*, Annex 1, June 9, 1977.

111. Kironde, "Rent Control Legislation and the National Housing Corporation in Tanzania, 1985–1990" 316.

112. Brennan, *Taifa*, 189.

113. Brennan, *Taifa*, 189.

114. Clearly many of these people also rented and did not build themselves.

115. Lawrence Kilimwiko, "Opting for Burnt Bricks," *Daily News*, July 26, 1982.

116. Askew, "Sung and Unsung," 28.

117. Forty, *Concrete and Culture*, 30.

118. Ingold, *Perception of the Environment*, 172.

119. Ingold, *Perception of the Environment*, 179.

120. Ingold, *Perception of the Environment*, 180.

121. Ingold, *Perception of the Environment*, 188.

Chapter 4: Waiting

1. "Disturbed Sleeper," *Daily News*, August 25, 1978. For more on the ubiquity of Radio Tanzania Dar es Salaam and its importance, see Perullo, *Live from Dar es Salaam*.

2. "Amka, Amka," *Daily News*, September 1, 1978.

3. "Leave Out the Bangs," *Daily News*, September 12, 1978.

4. Smith, "We Must Run While They Walk," *New Yorker*, October 30, 1971.

5. While loitering played a crucial role in Nyerere's conceptualization of ujamaa, laws against loitering have colonial origins. See Burton, *African Underclass*.

6. Quayson, *Oxford Street, Accra*, 241.

7. The long political life of loitering in colonial and postcolonial Tanzanian has been addressed by several Tanzanian scholars including Andrew Burton and James Brennan. See Burton, "Haven of Peace Purged," 119–51. Brennan, "Blood Enemies," 389–413.

8. Douglas, *Purity and Danger*.

9. African women were removed from urban spaces frequently through accusations of prostitution, a practice I do not discuss here but see as a gendered counterpart to loitering. For work on women's urban mobility, see Barnes, "Fight for Control of African Women's Mobility in Colonial Zimbabwe, 1900–1939," 586–608, and White, *Comforts of Home*.

10. This was implemented in 1923. Burton, *African Underclass*, 78.

11. Burton, *African Underclass*, 166.

12. On efforts to maintain colonial categories that were always vulnerable to dissolution, see Stoler, *Along the Archival Grain*.

13. TNA, File No 443 Provincial Office Eastern Province Folder: Traveling Permits and Passes for Natives," "Letter to the Acting District Officer of Dar es Salaam, 17th January 1939 from the Provincial Commissioner." 1938.

14. For more on this topic, see the next chapter, as well as Tripp, *Changing the Rules*; Maliyamkono and Bagachwa, *Second Economy in Tanzania*. For the control of such commerce in Nairobi, see Robertson, *Trouble Showed the Way*.

15. Burton, "Haven of Peace Purged," 121.

16. Burton, *African Underclass*, 1.

17. "Idlers are Enemies," *Daily News*, July 4, 1976, contains excerpts from Nyerere's speech declaring "war on drunkards, lazy people and other parasites" at Diamond Jubilee Hall, Dar es Salaam, June 25, 1976.

18. Molony, *Nyerere*, 163–65. While Nyerere's development may be unique, there is nothing exceptional about postcolonial condemnation of loitering and parasitism.

19. Nyerere, "Arusha Declaration" in *Ujamaa: Essays on Socialism*, 24.

20. See, for example, Latham, *Right Kind of Revolution*; and Thomas, "Modernity's Failings, Political Claims, and Intermediate Concepts," 727–40.

21. Rostow, *Stages of Economic Growth*; Williams, *Keywords*, 102–4. See also Cooper and Packard, *International Development and the Social Sciences*.

22. I am following Craig Jeffrey in his connection of personal acts of waiting to constructions of linear national development narratives. Jeffrey, *Timepass*, 12.

23. Chakrabarty, *Provincializing Europe*. See also Chakrabarty, "Muddle of Modernity," 663–75.

24. Mwakikagile, *Nyerere and Africa*, 440.

25. Alan Cowell, "Far from Tanzania, Nyerere Looks His Best," *New York Times*, March 14, 1982.

26. For example, "überholen ohne einzuholen," the GDR's slogan in the 1960s, means to "overtake without catching up."

27. "Idlers Are Enemies," *Daily News*, July 4, 1976.

28. This is despite the fact that Nyerere and Vice President Kawawa had condemned raids against loiterers during the colonial period. Burton, "Haven of Peace Purged," 121.

29. "Enforcement of Law on Work in Tanzania," *BBC Summary of World Broadcasts*, October 17, 1983.

30. Nyerere, *Five Years of CCM Government*.

31. Some self-employed workers and those who could prove they were farming could also remain.

32. "Letter to the Editor," *Daily News*, November 10, 1983.

33. Mororgoro Road, Pugu Road, Kilwa Road, and Bagamoyo Road (which branches off into "old" and "new" Bagamoyo Roads).

34. Burton, Brennan, and Lawi, *Dar es Salaam*, 53–54.

35. Project Planning Associates, *National Capital Master Plan, Dar es Salaam*, 67.

36. There were a variety of driving bans imposed after the oil embargo in 1973 to try to curb the amount of foreign exchange set aside for purchasing fuel from overseas.

37. Armstrong, "Colonial and Neocolonial Urban Planning," 43–66.

38. Armstrong, "Colonial and Neocolonial Urban Planning," 43–66.

39. Kulaba, *Housing, Socialism, and National Development in Tanzania. Bitumen* is the term used outside of the United States for tarmac.

40. Felix Kaiza, "Rebuilding Dar's Roads," *Daily News*, September 16, 1976.

41. "It's Time We Had Some Proper Planning," *Daily News*, February 17, 1973. Road building was the responsibility of the town planning division and the city council engineer is responsible for the maintenance of these roads (though the city council was abandoned in 1974).

42. Areas set aside for sites and services upgrading with the World Bank in 1974 and again in 1977 lacked paved roads, and in order to pave them, they would need to destroy about 8 percent of the area's current housing to provide enough room.

43. Project Planning Associates, *National Capital Master Plan, Dar es Salaam,* 36.

44. Felix Kaiza, "Rebuilding Dar's Roads," *Daily News*, September 16, 1976.

45. Two scholars of urban Africa who have written extensively on the opportunistic nature of informality are Filip De Boeck and AbdouMaliq Simone. See De Boeck and Plissart, *Kinshasa.* Simone, "Pirate Towns," 357–70.

46. Marshall, Macklin and Monaghan, Ltd., *Dar es Salaam Master Plan,* 26.

47. Briggs and Mwamfupe, "Peri-urban Development in an Era of Structural Adjustment in Africa: The City of Dar Es Salaam, Tanzania," 804. The British Dar es Salaam Motor Transport Company operated as a public transport monopoly in Dar from 1947 until it was nationalized in 1970. It remained the sole official provider of public transport until 1983. The DMT in 1970 was nationalized as part of the National Transport Corporation and the City Council. This transition to state ownership was framed as a key act in attaining Tanzania's socialist goals. "Special Meeting Held in the council chamber, Karimjee Hall, 28th May 1970," at 4:00 p.m., *City Council of Dar es Salaam Minutes 1970,* printed by Council's Rotaprint Town Clerk's Office, DSM, 316, Tanzania National Library. In 1974, the parastatal was divided in two companies to create an interregional transportation to Kampani ya Mabasi ya Taifa (KAMATA) while Dar's transport company became Shirika la Usafiri Dar es Salaam (UDA). While not discussed here, KAMATA suffered from many similar problems as UDA. See "Kampuni ya mabasi ya taifa," *Nchi Yetu,* September 1975.

48. Rizzo, "'Life Is War,'" 1183. For more history on public transport in Dar and its transformation under noeliberalism, see Rizzo's recent book, *Taken for a Ride* (2017).

49. Stren, "Administration of Urban Services," 51. This statistic is from 1984.

50. "DMT Machinery Needs Over Hauling," *Daily News*, October 7, 1973.

51. For example, the UDA in 1976 celebrating Union Day (the union of Tanganyika and Zanzibar) had the headline "All Out to Serve the Workers in Dar," and

featured a variety of newly purchased buses. In some socialist countries this was the case not just with buses but also shared cars. See Siegelbaum, *Socialist Car.*

52. Grace, "Modernization *Bubu*," 240.

53. By 1975, Freyhold argues that illiteracy in factories should have been wiped out and that it was "not unusual to see workers carrying around books at their work places and writing exercises during their work breaks. A worker found to be illiterate is in for heavy teasing from his colleagues." Freyhold, "Notes on Tanzanian Industrial Workers," 20.

54. Emily Callaci notes how central mobility, particularly male mobility, was to the image of "Afro-modernity." Callaci, "Street Textuality," 201. For more on masculinity, mobility, and modernity see chapter 3 in Hart, *Ghana on the Go.*

55. "Curb Bus Cramming," *Daily News*, October 22, 1973. "Frightening Facts on Tanzania's Motor Hazards," *Daily News*, January 20, 1974.

56. "Fuel Wastage in Dar," *Daily News*, March 28, 1974.

57. Richard Mngazija, "DMT Vipi?," *Nchi Yetu*, October 1973.

58. "Dar Roads Worsening," *Daily News*, January 26, 1984. Chiku Abdallah, "Kampuni ya Mabasi ya Taifa," *Nchi Yetu*, September 1975.

59. Salim Said Salim, "Special Sunday Interview on UDA Services," *Daily News*, June 5, 1977.

60. "DMT Machinery Needs Overhauling," *Daily News*, October 7, 1973. This experience struck locals and visitors alike and also could easily reach across class: even a visiting professor of math at UDSM made a point in reporting on his time in Dar to illustrate the horrors of trying to get around town. He recalled the head of the math department taking a bus into town only to have it break down "so he had to transfer to another bus after a suitable delay. This other bus also broke down before he got the nine miles into the centre of Dar es Salaam! So that a bus into Dar es Salaam would cost at least 1¼ hours and usually much more than this for one way." UKNA File No BW 90/1366: J E Phythian Professor of Mathematics "University of Dar es Salaam Visits by External Examiners Visit to Dar es Salaam, November-December 1974.

61. Rehema Shabani, interview with the author, Dar es Salaam, Tanzania, October 2009. See also "Insecure Bus Stops," *Daily News*, January 16, 1974.

62. Ivaska, *Cultured States*, 99.

63. *Mwongozo* was also the first major policy document not authored by Nyerere himself, but by Kingunge Ngombale-Mwiru (who was the secretary-general of the Tanzanian Youth League), with help from Rashidi Kawawa and Abdulrahman Babu. Roberts, "Politics, Decolonization and the Cold War in Dar es Salaam 1965–1972," 180.

64. Shivji, *Class Struggles in Tanzania*, 126.

65. Goran Hyden has compared *Mwongozo*'s influence on the lives of workers to the impact of the Arusha Declaration on the lives of Tanzania's peasants: "A promise of a better life with no specific conditions attached." Hyden, *Beyond Ujamaa in Tanzania*, 60.

66. TANU, *Mwongozo wa TANU*, 5.

67. Freyhold, "Notes on Tanzanian Industrial Workers," 15.

68. Hyden, *Beyond Ujamaa*, 160–67.

69. Freyhold, "Notes on Tanzanian Industrial Workers," 21.

70. "Between February 1971 and September 1973, there were 31 'downing of tools' involving something like 28,708 workers with 63,646 man-days lost." Shivji, *Class Struggles*, 136. See also Loxley and Saul, "Multinationals, Workers and the Parastatals in Tanzania," 83. Mihyo, "Struggle for Workers' Control in Tanzania," 62–84.

71. Shivji., *Class Struggles in Tanzania*, 127.

72. Luchiba Ngayile, "Too Much Noise against DMT," *Daily News*, February 24, 1973.

73. Shivji, *Class Struggles*, 141.

74. Mihyo, "Struggle for Workers' Control in Tanzania," 63–64.

75. Hyden, *Beyond Ujamaa*, 161.

76. Ironically, in 1978 when prime minister Sokoine visited the textile mill to celebrate its tenth anniversary, a spokesman for the workers took him aside to plead for private buses for the factory's four thousand employees. Sokoine complied, noting that all public institutions should have their own buses. Konde, *Press Freedom in Tanzania*, 133.

77. *Daily News*, August 11, 1974, as quoted in Konde, *Press Freedom in Tanzania*, 131–33, calling this divide an "anachronism" was expressing an idealistic view of parity within towns.

78. Salim Said Salim, "Special Sunday Interview on UDA Services," *Daily News*, June 5, 1977.

79. Stren, "Administration of Urban Services," 52.

80. "Firms Must Resume City Bus Service," *Daily News*, July 25, 1983.

81. From 1970 to 1974, they were referred to as Thumni Thumni, in reference to the fifty-cent coin it took to ride one, that name changed to Sanya Sanya which means to collect, gather, or steal from any source available.

82. Mohammed Halfani, interview with the author, Makongo Juu, Dar es Salaam, Tanzania, June 13, 2014.

83. The five-shilling coin was roughly equivalent to one US dollar at the time. Rizzo, "Being Taken for a Ride," 155.

84. Mihyo, "Struggle for Workers' Control in Tanzania," 65.

85. Freyhold, "Notes on Tanzanian Industrial Workers," 21.

86. Larkin, *Signal and Noise*, 247.

87. For work on "cascading effects" see Graham and Thrift, "Out of Order," 1–25. Little, "Controlling Cascading Failure," 109–23.

88. "Idlers are Enemies," *Daily News*, July 4, 1976.

89. In 1978 the UDA as well as KAMATA (the national bus service company) brought in six West German engineers as well as the two Hungarian experts already

working on Ikarus buses to rehabilitate ninety-four grounded buses (which were made by Ikarus, Leyland, Mercedes, and Fiat). The West German assistance program also covered the provision of spare parts and garage equipment. "UDA, KAMATA to Repair Buses," *Daily News*, January 10, 1978.

90. "UDA Still Needs Spares," *Daily News*, September 1976. *Fufua* in Swahili means to revive, resurrect, or bring back to life.

91. "We Lost 17 M, Says UDA," *Daily News*, May 3, 1982.

92. "We Lost 17 M, Says UDA," *Daily News*, May 3, 1982.

93. Hoyle, "African Politics and Port Expansion at Dar es Salaam," 39.

94. Veal, *Fela*, 122.

95. Veal, *Fela*, 123.

96. The port additionally served Zaire, Burundi, and Rwanda. After Southern Rhodesia announced its "unilateral independence" from Britain in 1965, Zambia's president Kenneth Kaunda and Nyerere (in an act of mutual solidarity against white minority rule in Southern Africa), were determined to create a new route for Zambia's access to international trade both for imported goods and for its lucrative copper industry. This led to the construction of the TAZARA railway, financed by the Chinese. For more on this see Monson, *Africa's Freedom Railway*.

97. "Importers Cause of Dar Port Congestion," *Daily News*, February 6, 1976.

98. The Dar port was also part of regional competition between Kenya, where the busier Mombasa port threatened to undermine Dar's feasibility. In the 1970s, Tanzania restricted truck traffic from Kenya to Zambia, concerned as it was about trade bypassing its own port. Hazlewood, *Economic Integration*.

99. For more on Breakdown, see Jackson, "Rethinking Repair."

100. Larkin, *Signal and Noise*, 219.

101. Larkin, *Signal and Noise*, 235–36.

102. For an examination of repair and "tinkering" and its implications for how we talk about "development," see Grace, "Modernization *Bubu*."

103. Grace, "Modernization *Bubu*," 224.

104. Simone, "People as Infrastructure," 407–29.

105. Derrida, "La différance," 73–101.

106. "Rains Cause Chaos on Dar Roads," *Tanganyika Standard*, February 29, 1968.

107. Temple, "Aspects of the Geomorphology of the Dar es Salaam Area," 39.

108. Stren, Halfani, and Malombe, "Coping with Urbanization and Urban Policy," 192.

109. Stren, "Administration of Urban Services," 52.

110. "UDA Drops Threat," *Daily News*, February 13, 1978. Attillio Tagalile, "Jogging to Get on a Bus," *Daily News*, May 10, 1983.

111. "Bus Shortage Hits Workers," *Daily News*, January 15, 1974. See also Richard Mngazija, "DMT Vipi?," *Nchi Yetu*, October 1973.

112. Stren, "Administration of Urban Services," 52.

113. For more on walking, see Porter, "Living in a Walking World," 285–300. Road safety and road accidents were a big issue in the 1970s: "Road Accidents Must Be Studied," *Daily News*, April 3, 1977. "Speeding Drivers Are a Danger," *Daily News*, April 3, 1977. "All Out War Against Accidents," *Daily News*, April 15, 1977.

114. Certeau, *Practice of Everyday Life*, 93.

115. Quayson, *Oxford Street, Accra*, 16.

116. Felix Kaiza, "Rebuilding Dar's Roads," *Daily News*, September 16, 1976.

117. Attillio Tagalile, "Jogging to Get on a Bus," *Daily News*, May 10, 1983.

118. Quayson, *Oxford Street, Accra*, 14.

119. Attillio Tagalile, "Jogging to Get on a Bus," *Daily News*, May 10, 1983.

120. Bienefeld, "Long-Term Housing Policy for Tanzania," 17.

121. L. Joel, "Like a Devil on a Mountain," *Daily News*, December 14, 1976.

122. Simone, "Waiting in African Cities," 97–109.

123. There has been some more recent anthropological work on waiting in Africa that suggests waiting in the neoliberal era is not marked by anticipation of the fruits of liberation but a sense of hopelessness of having nothing but time, particularly for young men. For more on this see Ralph's fantastic article "Killing Time," 1–29. Also, the chapter "Pumping Irony" in Quayson, *Oxford Street, Accra*. Similarly, Craig Jeffrey notes the term "timepass" as an Indian idiomatic for waiting. As Jeffrey explains, it is the result of "living in the shards of global capitalism." It is not passively experienced but rather "offered opportunities to acquire skills, fashion new cultural styles and mobilize politically" Jeffrey, *Timepass*, 4. He builds on Katz, "On the Grounds of Globalization," 1213–34.

Chapter 5: Wasting and Wanting

Epigraph: Untitled article, *Guardian* (UK), July 6, 1973.

1. "Dar Valleys for Distribution," *Daily News*, January 15, 1985. "Cultivate Dar Valleys—Hamad," *Daily News*, February 23, 1985.

2. "Re-think on Cultivating Dar's Valleys," *Daily News*, January 18, 1985.

3. Dan Malinga, "Preserve Dar Valleys," *Daily News*, February 19, 1985.

4. "Re-think on Cultivating Dar's Valleys," *Daily News*, January 18, 1985.

5. Sir Alexander Gibb and Partners, *Plan for Dar es Salaam*, 14. A copy of the master plan can be found in the Tanzania National Archives. In 1948, officials sought to fix the ailing system. The colonial secretary warned against more delays, arguing that "the built-up areas of the Township have become progressively more cesspit-riddled and sewage-sick." The community was in danger of contracting waterborne diseases, noted the secretary, "from which the only satisfactory safeguard is the installation of a system of water-borne sewerage and storm-water drainage." "Dar es Salaam Drainage," March 20, 1941, box 10 168, file #39/15, folder "Sewerage Scheme Dar es Salaam," acc: 450, TNA. By the 1970s, the city's decrepit sewage system was slated for improvement in the second master plan, but had to be upgraded leaving the system constantly vulnerable to "the rapid expansion of residential and industrial areas" and

creating "serious potential health hazards for the population." *Dar es Salaam Master Plan Summary*, 3. Copies of this master plan are located in several repositories in Dar, including the library in the Ministry of Lands, Housing, and Urban Development, DSM.

6. Armstrong, "Colonial and Neocolonial Urban Planning," 46.

7. In more recent years, the local government forbid farming along the creek due to pollution. For a more recent account of urban farming in Msimbazi, see McLees, "Access to Land for Urban Farming in Dar es Salaam, Tanzania," 601–24.

8. The distinction between first and second nature originates with Karl Marx. See Foster, *Marx's Ecology*. See also Cronon's introduction to *Nature's Metropolis* for more on the distinction between first and second nature, as well as Gandy, "Vicissitudes of Nature," 178–84.

9. Cronon, *Nature's Metropolis*. Williams, *Country and the City*. Tarr, *Search for the Ultimate Sink*. French, *When They Hid the Fire*. Reid, *Paris Sewers and Sewermen*. Gandy, *Concrete and Clay*.

10. Lofchie, "Agrarian Crisis," 451–75.

11. Lofchie, "Agrarian Crisis," 452.

12. Farmer cooperatives were nationalized in 1967 and then became relatively powerless. They were then abolished in 1976.

13. Lofchie, *Political Economy of Tanzania*, 77.

14. Lofchie, "Political and Economic Origins of African Hunger," 555.

15. Lofchie, "Agrarian Crisis," 457.

16. The collapse of agricultural production in Tanzania and across Tanzania has generated an extensive literature. See Ponte, *Farmers and Markets in Tanzania*. Bates, *Markets and States in Tropical Africa*. Lofchie, *Political Economy of Tanzania*.

17. Tanzania's foreign reserve holdings in June of 1973 were at 1.7 billion shillings. They dropped to 123 million shillings in November of 1974 and were so "uncomfortably low" in 1975 and 1976 that "even the most essential items could only be imported with great difficulty." Briggs, "Villagisation and the 1974–6 Economic Crisis in Tanzania," 696.

18. Masembejo and Tumsiph, *Upgrading in Dar es Salaam*, 10. As Shipton and others have pointed out, "the great irony of famines is that it is rural food producers who most often go hungry." Shipton, "African Famines and Food Security," 353–94.

19. Lofchie, "Agrarian Crisis," 458.

20. Goran Hyden examines how peasants opt out of the formal economy and turn to an "economy of affection," organizing economic activity through family, community, and ties of patronage rather than the formal economy of the state. See Hyden, *Beyond Ujamaa in Tanzania*.

21. Lofchie, *Political Economy of Tanzania*, 38–39.

22. Bryceson, "Century of Food Supply in Dar es Salaam," 171.

23. Mwana wa Matonya, "Where Shops are Empty Shelves," *Daily News*, January 28, 1973.

24. Ergas, "Why Did the Ujamaa Village Policy Fail?" 392–93. In the Dodoma region, "peasant families were ordered to grow at least three hectares of food crops, and if a man had a number of wives he must grow three hectares for every wife he had!" *Africa Contemporary Record* 1974, B287.

25. Bryceson, *Food Insecurity and the Social Division of Labour in Tanzania, 1919–85*, 215.

26. Lofchie, *Political Economy of Tanzania*, 75–76.

27. Alphonce Kyessi, interview with the author, University of Dar es Salaam, June 2014.

28. Fatuma (no last name given), interview with the author, Mbagala, Tanzania, October 2009. Others added that sometimes people mistakenly chased "dead body trucks" (i.e., those shuttling the deceased to burial sites), thinking they were food trucks. Mwenevyale (no last name given), interview with the author, Mbagala, Tanzania, October 21, 2009. Mohamedi Mikoi, interview with the author, Mbagala, October 2009.

29. Maliyamkono and Bagachwa, *Second Economy in Tanzania*, 93.

30. "Survey of the Kisutu Market, Dar es Salaam," Marketing Development Bureau FAO-UNDP Project SF TAN 27 Ministry of Agriculture August 1972, 18. Available at Herskovitz Library, Northwestern University, USA.

31. For more on produce supply in the city, see Lynch, "Urban Fruit and Vegetable Supply in Dar es Salaam," 307–18.

32. Rahema Shabani, interview with the author, Mbagala, Tanzania, October 23, 2009.

33. Shabani Omari Mlanzi, interview with the author, Mbagala, October 2009.

34. Nyerere, *President Nyerere's Speech to Parliament, 18th July, 1975*, 9.

35. For more on women as urban cultivators, see Freeman, "Survival Strategy or Business Training Ground?" 1–22. Lee-Smith and Trujillo, "Struggle to Legitimize Subsistence," 77–84.

36. Mclees notes that there is a distinction now in Dar between "home gardens," typically tended by women and "open space" farms, which tend to be farmed by men. "Open space farming differs from home gardens because several farmers work at one farm and each farmer has several plots, often depending on the number of years he or she has spent farming in the area. In Dar es Salaam, open-space cultivation occurs [today] on policy land, on school properties, on road reserves, under powerlines, on private company land and on university land." McLees, "Urban Farming in Dar es Salaam," 603.

37. Hamada Ali Mnora, interview with the author, Mbagala, November 2009.

38. A survey of female wage earners in 1973 noted that 67 percent of the women surveyed sold fruits and vegetables that they had picked up at markets or from their own small farms. Swantz and Bryceson, *Women Workers in Dar es Salaam*.

39. For more on the interplay between urban agriculture, form and informal work see chapter 2 in Tripp, *Changing the Rules*.

40. Swantz and Bryceson, *Women Workers in Dar es Salaam*.

41. Tripp, "Urban Farming and Changing Rural-Urban Interactions in Tanzania," 105.

42. Bryceson, *Food Insecurity in Tanzania*, 214.

43. Tripp notes that by 1987, 87 percent of those who had left employment in the city for reasons other than retirement had gone into self-employment and 40 percent had gone into urban farming. Of those who remained employed, 38 percent planned or hoped to leave their jobs to engage in small enterprises or farming or both. Tripp, "Defending the Right to Subsist."

44. Lofchie, *Political Economy*, 85.

45. "Mwalimu Tours Dar," *Daily News*, June 1976.

46. See Lofchie, *Political Economy*, 84–87.

47. Chachage, "Forms of Accumulation, Agriculture and Structural Adjustment in Tanzania," 199.

48. "Dar Residents Urged to Farm," *Daily News*, August 27, 1985.

49. Attilio Tagalilie, "Dar Peasants Left on their Own," *Daily News*, March 16, 1985.

50. For more on this shift to the suburbs and "ruralization" of the city, see Mercer, "Landscapes of Extended Ruralisation," 72–83. Briggs, "The Peri-urban Zone of Dar es Salaam, Tanzania," 319–31. Owens, "Post-colonial Migration," 249–74.

51. By 1978: 51,000 counting within the urban planning laws and 32,000 in rural planning areas, Marshall, Macklin and Monaghan, Ltd., *Dar es Salaam Master Plan Summary*, 26.

52. Mbilinyi and Mascarenhas, *Sources and Marketing of Cooking Bananas in Tanzania*.

53. Mkumbwa Ally, "Dar Residents to Get Farms," *Daily News*, June 19, 1983.

54. Daniel Mshana, "City Expansion Plan Approved," *Daily News*, December 25, 1982. See also: Sawio, *Urban Agriculture and the Sustainable Dar Es Salaam Project*. Number of dairy cattle in Dar is 4,200, more than doubles by 1989, and goats triple from 2217 to 6218 and pigs double from 8601 to 15658 chickens double from 500,000 to 1,000,000.

55. Donna Kerner, "'Hard Work' and Informal Trade Sector in Tanzania," 48.

56. "City Out to Clear Roads," *Daily News*, December 5, 1979.

57. See Brownell, "Seeing Dirt in Dar es Salaam."

58. The dump was finally closed due to the efforts of a community group known as the Tabata Development Fund. See Kessy, "Promoting Good Urban Governance at the Community Level," 133–36.

59. Marshall, Macklin and Monaghan, Ltd., *Dar es Salaam Master Plan Technical Supplement Sewage Collection and Disposal*, 158, and *Daily News*, March 15, 1974.

60. Marshall Macklin Monaghan, *Dar es Salaam Master Plan Technical Supplement Sewage Collection and Disposal*, 158. Dar es Salaam was also a port city and the major sight for industry in Tanzania and yet these industrial waste streams are hard

to track down and are rarely discussed in newspapers or much in the three generations of master plans.

61. Felix Kaiza, "Something Rotten in Town," *Daily News*, July 8, 1977.

62. "Keep the City Clean," *Daily News*, January 6, 1976.

63. "Keep Towns Clean Says Mwalimu," *Daily News*, February 11, 1983.

64. "Campaign on Filth Starts in Dar," *Daily News*, January 5, 1978.

65. "UWT Members Clean-Up Club," *Standard*, March 7, 1968. "Campaign on Filth Starts in Dar," *Daily News*, January 5, 1978.

66. Mtambo J.P. Mtambo, "City Cleanliness Campaign," *Daily News*, November 28, 1985.

67. "Cholera Kills 160 in 3 Months," *Daily News*, January 17, 1978. "Cholera Claims More Lives," *Daily News*, January 25, 1978. "Dar is Still Dirty," *Daily News*, March 21, 1978.

68. John A. O. Max in his book, *The Development of Local Government in Tanzania* (1991), 90, also notes that a 1976 cholera outbreak had prompted a study regarding whether urban councils should be reinstated. Their reinstatement, however, did not happen until after the outbreak in 1978. Nyerere initiated a Presidential Commission in 1978 to examine the problems facing Dar and in July of 1978, urban councils across Tanzania were reinstated. Nyerere later in 1985 claimed that the dissolution of urban councils had been a "grave mistake which would never be repeated." *Daily News*, March 30, 1985. See Paddison, "Ideology and Urban Primacy in Tanzania," 19.

69. Dar's sewage system initially covered the commercial area of the city and a small part of Kariakoo. "It was later extended to Muhimbili and Upanga, converging at Ocean Road from where it proceeds to empty into the sea." See the Project Report for "Waste Disposal in Dar es Salaam," 4. Can be found in the East Africana Library, University of Dar es Salaam.

70. Frank E. Jones, "Operation and Maintenance Services for Dar es Salaam Sewerage and Sanitation."

71. World Bank, "Staff Appraisal Report," 3.

72. World Bank, "Staff Appraisal Report," 8–9.

73. World Bank, "Staff Appraisal Report," 8–9.

74. See also Brownell, "Seeing Dirt in Dar es Salaam."

75. Nyerere, "Arusha Declaration Ten Years After," 38.

76. See Gille, *From the Cult of Waste to the Trash Heap of History*.

77. Men and boys in particular also found employment in recovering waste:

> Boys at Hargeisa in Somaliland have organised themselves into an engineering firm on a rubbish dump. In this way from twenty-five to thirty boys support themselves. They live, rent free, in a disused shed, and grew four sacks of millet on a piece of land lent them for the purpose. Whereas the very poor of Yoruba towns had scoured the bush, the very poor of colonial towns scavenged industrial wastelands. Sanitary workers in Lourenco Marques reworked collected trash and resold bottles, plastic bags, rope, metal, old clothes, and a host of other

articles. Ibadan had an Association of Worn Out Tyre Traders. In Abidgan men toured the streets with bathroom scales offering to weigh people for two pence a time. (Iliffe, *African Poor*, 175)

78. Strasser, *Waste and Want*, 22. I also riff on Strasser for the title of this chapter.

79. Lofchie, *Political Economy*, 29.

80. These were not the only neighborhoods in town that became home to small industries. A *Daily News* article from November 8, 1978, about "Manseze's quiet revolution," notes that the neighborhood once considered a scourge on the city was now home to "garages, carpentry workshops and makers of household utensils from scrap metal, whose incomes vary from a few hundred shillings to 6000 a month are springing up."

81. Kent and Mushi, "Education and Training of Artisans," 85.

82. Before 1973, it was known as the National Self-Industries Corporation or the National Small Industries Corporation.

83. Kent and Mushi, "Education and Training of Artisans," 87.

84. Virtually no women were members. From 1979 to 1992, one woman was involved in paper bag manufacturing and two others tried but had to drop out due to objections from their parents.

85. Livengood, "Mafundi Chuma and Folk Recycling in Dar Es Salaam," 87.

86. Yhdego, "Scavenging Solid Wastes," 263.

87. Yhdego interviewed Takataka in 1990.

88. Yhdego, "Scavenging Solid Wastes," 262.

89. Yhdego, "Scavenging Solid Wastes," 262.

90. Kent and Mushi, "Education and Training of Artisans," 87.

91. Kent and Mushi, "Education and Training of Artisans," 143.

92. Havenvik, Skarstein, and Wangwe, *Small Scale Industrial Sector Survey*.

93. Havenvik, Skarstein, and Wangwe, *Small Scale Industrial Sector Survey*, 13.

94. Havenvik, Skarstein, and Wangwe, *Small Scale Industrial Sector Survey*, 21.

95. Perkins, "Technology Choice," 240.

96. Havenvik, Skarstein, and Wangwe, *Small Scale Industry Sector Survey*, 172.

97. Havenvik, Skarstein, and Wangwe, *Small Scale Industry Sector Survey*, 25.

98. I could not find out why. Perkins, "Technology Choice," 240.

99. This may be a problematic statistic due to incomplete surveying and the problem of categorizing formal and informal operations.

100. Nyerere's speech to the Institution of Engineers was reprinted as "The Tanzanian Engineer Must Be a Very Practical Person," *Daily News*, February 13, 1985. He declared that "Appropriate Technology will be that which emphasises the use of local materials, and local human resources such as they are now. It will thus use wood, sisal waste, and the by-products of existing operations, instead of imported raw materials: and it will call for unskilled or semi-skilled labour in operation rather than involving heavy capital costs. It will be tough though, so that it can withstand

rough use, especially when it is to be installed in a village or in the streets of our towns. It will be simple to operate and to repair; it should for preference be able to be repaired very near to where it is used."

101. In particular, scholars like Robert Bates have critiqued the heavy focus on import substitution at great cost to farmers, creating what he and others called "urban bias." Bates, *Markets and States in Tropical Africa*.

102. Lofchie notes that this decision was heavily critiqued: "The development economists were clear in their insistence that a country should not move forward to the second stage until it met these preconditions. Tanzanian leaders held a different view. As if to demonstrate their unflagging commitment to ISI they decided in 1976 to move on to the second stage, which they called the basic industries strategy." Lofchie, *Political Economy of Tanzania*, 101–2.

103. Nyerere, *Ujamaa*.

104. For a sense of scope, "from 1976–1981, India's share of Tanzania's import of non-electric machinery ranged from 4% to 12%" with its highest in 1979. In food machinery only, it was as much as 73% in 1978, 86% in 1979 and 56% in 1980. Folke, Fold, and Enevoldsen, *South-South Trade and Development*, 202.

105. Enevoldsen, Fischer, Fold, and Folke, *India's Export of Capital Goods to Tanzania, 15–17. Available at the British Library of* Development Studies, Sussex, UK.

106. Lall, *Multinationals, Technology and Exports*, 223.

107. "Technology is Not Neutral," *Daily News*, May 11, 1977.

108. "Review: The Barbed Wire by Mukotani Rugyendo," *Daily News*, August 19, 1978. Rugyendo was a Ugandan who graduated from the University of Dar es Salaam in 1973 and later served as an editor at the Tanzanian Publishing House in Dar es Salaam.

109. Enevoldsen, Fischer, Fold, and Folke, *India's Export of Capital Goods to Tanzania, 39.*

110. *Regional and Country Studies* Branch Division for Industrial Studies. *United Republic of Tanzania.*

111. Lofchie writes that Tanzania's industrial strategy was stymied because "the level of capital investment required was too great and industries producing capital goods were so complex that there would need to be a long-term donor commitment to provide the necessary financial resources. The new industries, in other words, required dependence on foreign assistance. By implementing the basic industry strategy, Tanzania had transformed the ISI approach, taking it from an economic program whose goal was industrial self-sufficiency into a purposeful means of extracting long-term financial support from the donor community." Lofchie, *Political Economy of Tanzania*, 103.

112. "Technology is Not Neutral," *Daily News*, May 11, 1977.

113. Folke, Fold, and Enevoldsen, *South-South Trade and Development*, 202.

114. Perkins, "Technology Choice, Industrialisation and Development Experiences in Tanzania," 231.

115. Havenvik, Skarstein, and Wangwe, *Small Scale Industrial Sector Survey,* 202.

116. Enevoldsen, Fischer, Fold, and Folke, *India's Export of Capital Goods to Tanzania, 81.*

117. Julius Nyerere, "The Tanzanian Engineer Must Be a Very Practical Person," *Daily News,* February 13, 1985.

118. Perkins, "Technology Choice, Industrialisation and Development Experiences in Tanzania," 241.

119. John Waluye, "Kibo Paper Appeals: Don't Burn Waste Paper," *Daily News,* March 9, 1985.

120. Pili Mtambalike, "Use Local Materials, Plastic Firms Told," *Daily News,* January 21, 1983.

121. Mtambalike, "Use Local Materials."

122. Waluye, "Plastic from Waste."

123. "ALAF Collects Scrap Metal," *Daily News,* April 20, 1982.

124. Daniel Mshana, "Industries Advised on Energy," *Daily News,* April 5, 1983.

125. "Interview with the Minister for Industries, ND. Basil Mramba," in IDO Special Supplement, *Daily News,* December 6, 1983.

126. Wence Mushi, "Mwalimu Stresses Self-Reliance," *Daily News,* October 21, 1982.

127. For more on this see Aminzade, *Race, Nation, and Citizenship in Post-Colonial Africa,* 246.

128. US National Archives, (US) Series: Country Files Subseries: Country Files Box: 3 File 1 Title C/Tanzania/701 Fund Assistance to and Relations with Members. Transcript of "President Nyerere's Speech at the Dinner Given for Diplomats Accredited to Tanzania, 1st January 1980 At the Kilimanjaro Hotel, Dar es Salaam."

129. Another common term for this was *unyonyagi,* or bloodsucking. See Brennan, "Blood Enemies," 387–411. A third term, for smuggling in particular, was *magendo,* also used in neighboring Kenya and Uganda.

130. Lofchie, *Political Economy,* 35.

131. Mkumbwa Ally, "Ministry Plans Strict Trade Supervision," *Daily News,* July 7, 1983.

132. For example, here are some official versus open market prices for consumer goods in Mbeya in 1984: Khanga (local) cloth unit: Official Price 200 Market Price 800; Washing Soap Unit: Block official price 15 Market Price 50; Cooking oil kg official price 80 Market price 200; Sugar kg official price 15 market price 50. January 30, 1985, Memorandum to: Mr. Anupam Basu Division Chief Mideast African Div., IMF from Florent Agueh, Chief EAISE, IMF African Department Fonds Immediate Office Sous-fonds Series: AFRAI Country Files Box 126 File 3 Tanzania—Correspondence 1985 January-March Office Memorandum.

133. "Graduates Hail Crackdown," *Daily News,* May 16, 1983.

134. As Lofchie notes in *Political Economy:* "Corruption was Tanzania's form of early capital accumulation" (24). In his memoir, former Minister of Finance Edwin

Mtei recalled how government officials used travel opportunities across state borders to accumulate scarce goods. "If an official travelled to Mbeya [near the border with Zambia and Malawi], he would return to Dar es Salaam with two bags of rice in his official vehicle. If he saw a transistor radio in a shop, he would buy it even though he might possess several already at home. Similarly with furniture: houses occupied by well-to-do individuals were congested with items of furniture acquired simply to avoid the escalating inflation." Mtei, *From Goatherd to Governor*, 165.

135. Mkumbwa Ally and Adam Lusekelo, "Scarce Items Resurfacing," *Daily News*, April 8, 1983.

136. See *Daily News*, April 1, 1983; April 7, 1983; April 8 1983; May 22, 1983.

137. Richard Hall, "Tanzania Hunts Hoarders of Cash, Goods," *Globe and Mail* (Canada), April 22, 1983.

138. "Tanzania: War on Saboteurs," *Africa Journal*, ltd. issues 137–148, 23.

139. "Crackdown Paving Way for Ujamaa," *Daily News*, April 12, 1983. For a good accounting of how saboteurs acquired goods according to the state, see "Sokoine Outlines Sabotage Tricks," *Daily News*, April 23, 1983.

140. "Foul Play Tales on Sokoine's Death," *Citizen*, April 12, 2016, accessed July 16, 2018, http://www.thecitizen.co.tz/News/Foul-play-tales-on-Sokoine-s-death /1840340-3155326-rt28g6z/index.html.

141. "Guiding On-going War," *Daily News*, May 20, 1983.

142. "Guiding On-going War," *Daily News*, May 20, 1983.

143. "Guiding On-going War," *Daily News*, May 20, 1983.

144. Maliyamkono and Bagachwa, *Second Economy in Tanzania*, xii.

145. Ally and Lusekelo, "Scarce Items Resurfacing," *Daily News*, April 8, 1983. "Surrender Goods, Money," *Daily News*, April 9, 1983.

146. "Comment" column, *Daily News*. June 7, 1983.

147. "War That Must Be Won," *Daily News*, May 11, 1983.

148. "Crackdown Paving Way for Ujamaa," *Daily News*, April 12, 1983.

149. "Dar Gets Water Again," *Daily News*, January 22, 1984.

150. At different decibels and to different ends, the language of rooting out both real and metaphorical dirt is a pan-African phenomenon in the postcolonial era. Alicia Decker has pointed out that a very same language about pollution, dirt, and cleanliness circulated nearby in Idi Amin's Uganda in the 1970s. Amid Amin's murderous efforts to retain power and root out any saboteurs of his regime, he also expended an immense amount of energy and political authority on an urban campaign called Operation Keep Uganda Clean. Decker argues that "dirt served as a powerful metaphor for subversion and sabotage—activities that threatened the regime's stability. For Amin, campaigns for cleanliness shored up his authority while also function "as a coded language for eliminating political dissent." Decker, "Idi Amin's Dirty War," 490–91. This happened in Zaire too, where Amin possibly took the idea. "Salongo" in Zaire was a broader compulsory state project nonetheless framed as "voluntary," that compelled food crop cultivation as well as compulsory afternoons cleaning bore-

holes or painting houses. There are examples of similar campaigns in Babangida's Nigeria as well during his "War against Indiscipline" in the early 1980s.

The political power of pointing out waste that the state so readily drew upon could also be subverted as a powerful tool of opposing authority as well. Postcolonial African writers have frequently employed a political rhetoric of waste and wasted people, or put more crudely, of shit. In his article on "excremental postcolonialism," Daniel Esty notes the prevalence of excrement as a "governing trope" in postcolonial African fiction for a generation of writers stuck between jubilation at the end of colonialism and the disillusionment of nationalist sentiment in the independence era. "Shit has a political vocation," Esty writes, "it draws attention to the failures of development, to the unkept promises not only of colonial modernizing regimes but of post-independence economic policy."(23) Writers employed it narratively and to serve figuratively as "a material sign of underdevelopment; as a symbol of excessive consumption; as an image of wasted political energies." Etsy, "Excremental Postcolonialism," 14. See also Anderson, "Crap on a Map, or Postcolonial Waste," 169–78.

151. For more on the question of nonscalability or descaling, see Tsing, "On Nonscalability," 505–24.

152. *Daily News*, September 26, 1983.

153. For a letter to the editor that captures this irony, see "License Street Vendors" by "Prono Publico," *Daily News*, July 16, 1985.

154. *Sunday Daily News*, October 2, 1983.

155. "World Environment Day Article: Which Urban Alternative for Africa?" *Daily News*, June 8, 1979.

Chapter 6: Fueling Crisis

1. Nkonoki, *Energy Crisis of the Poor in Tanzania*, 56.

2. National Research Council, *Firewood Crops*, 29.

3. Eckholm, "Other Energy Crisis: Firewood."

4. Mavunga, "*Cidades Esfumaçadas*," 270.

5. Shechambo, "Urban Demand for Charcoal in Tanzania," 1.

6. This notion of charcoal as an anarchic fuel comes both from Mitchell's notion of how the materiality of oil production and transport produced its politics "carbon democracy" and also Scott, *Against the Grain*, which argues that the materiality/seasonality, and nature of certain agricultural crops permitted communities to escape being governed while grain production increased state power over other communities.

7. Nearly one million hectares are classified as "closed forests" with the rest (32.6 million hectares) classified as open, less dense woodlands. Persson, *Forest Resources of Africa*, 74.

8. Sunseri, *Wielding the Ax*, 20, 2.

9. Sunseri, *Wielding the Ax*, 2.

10. In particular see the work of Thaddeus Sunseri cited throughout this chapter and Neumann, "Forest Rights, Privileges and Prohibitions," 45–68. Steinbach, "Carved Out of Nature." Andrew Hurst, "State Forestry and Spatial Scale," 358–369.

11. Sunseri, "Every African a Nationalist," 888.

12. *Forest Ordinance of Tanzania (1957)*.

13. Sunseri, "Every African a Nationalist," 885.

14. See Sunseri, "Something Else to Burn."

15. While many native species were cut down and replaced with nonnative trees, some original forests remain now marked as conservation areas such as Pugu Hills and Kazimzumbwi, today considered some of the oldest forests in the world.

16. Sunseri, "Something Else to Burn," 617.

17. Neumann, "Forest Rights," 60.

18. Sunseri, "Every African a Nationalist," 888.

19. Neumann, "Forest Rights," 63.

20. Sunseri, "Something Else to Burn," 623.

21. Accession 604: FD/39/20/14 Forest Industries Development Charcoal 1963–72, TNA

22. Sunseri, *Wielding the Ax*, 142.

23. It is also worth noting that charcoal has long been a product of Indian ocean trade between East Africa and the Arabian peninsula. Accession 604: FD/39/20/14 Forest Industries Development Charcoal 1963–72, TNA.

24. Report: "Tanzania Charcoal Manufacturers Limited (Tancoal) (in formation)," FD/39/20/14 Forest Industries Development Charcoal 1963–72 Accession #604: TNA.

25. See Openshaw and Food and Agriculture Organization, *Forest Industries Development Planning*.

26. FD/39/20/14 Forest Industries Development Charcoal 1963–72, Accession #604, TNA.

27. Ibid.

28. FD/39/20/14 Forest Industries Development Charcoal 1963–72, Accession #604, TNA.

29. Sunseri, *Wielding the Ax*, 159.

30. Nkonoki, *Energy Crisis of the Poor in Tanzania* Survey, annex VIII, 112–13.

31. Chaix, "Is a Charcoal Crisis Looming for Tanzania?" *Charcoal Project*, January 19, 2010, accessed August 2019, http://www.charcoalproject.org/2010/01/is-a-woodfuel-and-charcoal-crisis-looming-for-tanzania/.

32. Nkonoki, *Energy Crisis of the Poor in Tanzania*, 107.

33. Nkonoki, *Energy Crisis of the Poor in Tanzania*, 111.

34. Sunseri, "Something Else to Burn," 626.

35. Nkonoki, *Energy Crisis of the Poor in Tanzania*, 111.

36. Tripp, *Changing the Rules*, 135.

37. Nkonoki, *Energy Crisis of the Poor in Tanzania*, 122.

38. Nkonoki, *Energy Crisis of the Poor in Tanzania*, 108–9.

39. Nkonoki, *Energy Crisis of the Poor in Tanzania*, 109.

40. Mavhunga in his article on charcoal in Mozambique also makes this point. Mavhunga, "*Cidades Esfumaçadas*," 261–71.

41. Simone, "People as Infrastructure," 407.

42. Peter and Sander, *Environmental Crisis or Sustainable Development Opportunity?*

43. Nwgare, "Environmental Degradations and Fuelwood Consumption," 95.

44. Nwgare, "Environmental Degradations and Fuelwood Consumption," 82.

45. Nwgare, "Environmental Degradations and Fuelwood Consumption," 98.

46. Nwgare, "Environmental Degradations and Fuelwood Consumption," 92–93.

47. Nwgare, "Environmental Degradations and Fuelwood Consumption," 94.

48. Nwgare, "Environmental Degradations and Fuelwood Consumption," 98.

49. Darkoh, "Desertification in Tanzania," 329.

50. Meadows et al., *Limits to Growth*. The term "northern environmentalists" refers to the vocal policy advocates and scientists of the nascent global environmental discourse that emerged in Europe and the United States in the 1970s (for more on these individuals see Macekura, *Of Limits and Growth*. While it is surely reductionist to lump all "northern environmentalists" together with such a term, for my purposes here I feel it is appropriate. First, this is because there did emerge a notable consensus among environmentalists that there was an unfolding planetary crisis. Indeed, this broad consensus and scope were central to this new era of identifying "global" problems. Secondly, I believe that the influence of this new conception of a global environment generated by environmental advocates in the Global North was experienced by Tanzanians in a rather unnuanced way. It was not about individuals but the policies these powerful advocates for the environment were able to initiate. Finally, the term also reflects the rather lumpen way that "northern environmentalists" were often guilty of seeing the rest of the world, the threat of the developing world, and the redemptive power of forests. For more on the consensus among northern environmentalists, particularly about forests and deforestation, see Dove, "Forest Discourses in South and Southeast Asia: A Comparison with Global Discourses"; and McCann, "Plow and the Forest," 138–59.

51. Mitchell, *Carbon Democracy*; chapter 7, "The Crisis That Never Happened."

52. Mitchell, *Carbon Democracy*, 188.

53. Mitchell, *Carbon Democracy*, 189.

54. Caldwell, "World Policy for the Environment," 6.

55. Macekura, *Of Limits and Growth*, 93.

56. Macekura, *Of Limits and Growth*, 93.

57. Many of these activists also saw themselves as allies of the Third World. It is impossible and unnecessary here to draw clear lines of solidarity along divisions

of the First and Third World. There were, however, two general schools of thought regarding resource use, development, and new interventions in the name of conservation that do align along these contours.

58. Farvar and Milton, *Careless Technology.*

59. Gandhi, "Indira Gandhi's Speech at the Stockholm Conference in 1972."

60. Sachs, *Development Dictionary,* 27.

61. Davis, "Deserts," 117.

62. More recently it has become highly debated to what extent humans cause desertification. Herrmann and Hutchinson, "Changing Contexts of the Desertification Debate," 539.

63. Davis, "Deserts," 117.

64. Shechambo, "Urban Demand for Charcoal in Tanzania," 2.

65. Macekura, *Of Limits and Growth,* 129.

66. United Nations, *Resolution Adopted by the General Assembly 44/172 Plan of Action to Combat Desertification.*

67. Lester R. Brown, preface, in Eckholm, "Other Energy Crisis."

68. For a sense of the literature generated on just Tanzania see the references section in Johnsen, "Burning with Enthusiasm," 107–31.

69. Taylor, "How Do We Know We Have Global Environmental Problems?" 149.

70. Meadows, Meadows, Randers, and Behrens, *Limits to Growth.* In addition to Macekura's book, another recent book to look back at the importance of this book is Higgs, *Collision Course.*

71. Leach and Mearns, *Beyond the Woodfuel Crisis,* 6.

72. Leach and Mearns, *Beyond the Woodfuel Crisis,* 7.

73. Taylor, "How Do We Know We Have Global Environmental Problems?" 153. For more on modeling environmental problems, see Edwards, *Vast Machine.*

74. Taylor, "How Do We Know We Have Global Environmental Problems?" 146.

75. Taylor, "How Do We Know We Have Global Environmental Problems?" 152.

76. As Clapperton Mavhunga points out in his article, *"Cidades Esfumaçadas,"* 261–71. In Mozambique where an immense amount of energy is generated from the Cahora Bassa Dam, Maputo's residents still mostly use charcoal because 70 percent of the energy is sold to South Africa.

77. World Bank, *Tanzania: Issues and Options in the Energy Sector,* 30.

78. One example of how ubiquitous this pairing became is a 1984 Earthscan publication by Foley, Moss, and Timberlake, *Stoves and Trees.*

79. For example, in 1981 there was a UNEP conference held in Nairobi on alternative fuels. see *Report of the United Nations Conference on New and Renewable Sources of Energy, Nairobi, 10 to 21 August 1981.*

80. Schumacher, *Small Is Beautiful.*

81. Quoted in Johnsen "Burning with Enthusiasm," 124.

82. World Bank, *Tanzania: Issues and Options in the Energy Sector*, 128.

83. For more on how the solar cooking stove arose as one of the key objects of the Appropriate Technology movement beginning in the 1950s, see Pursell, "Appropriate Technology, Modernity, and U.S. Foreign Aid," 175–87.

84. World Bank, *Tanzania: Issues and Options in the Energy Sector*, 24.

85. World Bank, *Tanzania: Issues and Options in the Energy Sector*, 25.

86. World Bank, *Tanzania: Issues and Options in the Energy Sector*, 25.

87. World Bank, *Tanzania: Issues and Options in the Energy Sector*, 25.

88. Nkonoki, *Energy Crisis of the Poor in Tanzania*, 57.

89. World Bank,*Tanzania: Issues and Options in the Energy Sector*, 79.

90. Kammen, "Research, Developoment and Commercialization of the Kenya Ceramic Jiko and Other Improved Biomass Stoves in Africa."

91. Nash and Luttrell, "Crisis to Context," 2.

92. O'Keefe and Soussan, "Energy: Power to Some People," 111.

93. O'Keefe and Soussan, "Energy: Power to Some People," 111–12.

94. Wainaina, "Glory."

95. Kenya Woodfuel Development Programme, *So, Firewood Can Wreck a Home?*

96. Kenya Woodfuel Development Programme, *So, Firewood Can Wreck a Home?*

97. Hunt "'It's More Than Planting Trees, It's Planting Ideas,'" 239.

98. Schell, "Transnational Environmental Justice Rhetorics and the Green Belt Movement," 17.

99. "Dar Urged to Plant Trees," *Daily News*, March 27, 1985.

100. "Dar Urged to Plant Trees," *Daily News*, March 27, 1985.

101. Skutsch, "Why People Don't Plant Trees."

102. Taylor, "How Do We Know We Have Global Environmental Problems?" 157.

103. Persson, *Forest Resources of Africa*, 74.

104. Food and Agriculture Organization and United Nations Environment Programme, *Tropical Forest Resources Assessment Project*.

105. World Bank, *Tanzania: Issues and Options in the Energy Sector*, 23.

106. Skutsch, "Why People Don't Plant Trees," 74.

107. Mnzava, *Tree Planting in Tanzania*, 73.

108. Mnzava, *Tree Planting in Tanzania*, 73.

109. Skutsch, *Why People Don't Plant Trees*, 10.

110. Zollner, "Village Woodlot Project in Tanzania."

111. "18 Million Trees Planted," *Daily News*, May 24, 1985.

112. Mnzava, *Tree Planting in Tanzania*, 3–7.

113. *Nairobi Programme of Action for Woodfuels*, http://www.fao.org/docrep/t0747e/t0747e03.htm.

114. Mnzava, *Tree Planting in Tanzania*.

115. Peter and Sander, *Environmental Crisis or Sustainable Development Opportunity?*, 6. See also Nash and Luttrell, "Crisis to Context."

116. An example of Eckholm and Worldwatch seeking out woodfuel coverage: an article in the *Northwest Arkansas Times* from Fayetteville, Arkansas, in 1976 was written by a local reporter, Peggy Frizzell, who noted a recent visit to Fayetteville by a Worldwatch staff member and that after having read the "Other Fuel Crisis" the reporters "sat around, [drank] coffee and talked about the firewood situation in Nepal" Frizzell, "In Nepal, They Burn Cow Dung," *Northwest Arkansas Times*, November 1, 1976.

117. See "Nyerere Urges Tree Planting Campaigns," *Daily News*, August 21, 1980; "Use of Charcoal on the Rise Five Times," *Daily News*, September 21, 1980; "Firewood: The Other Energy Crisis," *Daily News*, May 12, 1982; "Afforestation Makes Progress," *Daily News*, May 12, 1981; "Forest Campaign Doing Well," *Daily News*, May 25, 1981; "Tree Planting Gains Tempo," *Daily News*, June 1, 1981.

118. Schell, "Transnational Environmental Justice Rhetorics and the Green Belt Movement," 585–613.

119. Cohen, *Planting Nature*, 2.

120. Schell, "Transnational Environmental Justice Rhetorics and the Green Belt Movement," 606.

121. Within American history, tree-planting was used to "settle" and domesticate arid regions of the West, reinventing them in ecologically problematic ways. In North Africa, tree planting was used to justify colonial occupation by invoking the practice as a restoration of the region's mythical green past. See Davis, *Resurrecting the Granary of Rome*.

The engineers of Israeli's Afforestation Project also used eucalyptus trees to quickly cover over inconvenient ruins of destroyed Palestinian villages. Called "security groves," these fast-growing trees replaced olive orchards with the intention of also erasing a history of settlement and ownership. See Cohen, "Politics of Planting."

122. Deleuze and Guattari, *Thousand Plateaus*. Malkki, "National Geographic," 24–44. Wampole, *Rootedness*.

123. For more on the rise of NGOs, particularly in Tanzania see Jennings, *Surrogates of the State*.

124. Macekura, *Of Limits and Growth*, 99.

125. For Tanzania specifically see Jennings, *Surrogates of the State*. For a broader discussion of this see Ferguson, *Anti-politics Machine*.

126. O'Keefe and Soussan, "Energy," 113.

127. O'Keefe and Soussan, "Energy," 113.

128. Aminzade, *Race, Nation, and Citizenship in Post-colonial Africa*, 246.

129. Bagachwa, "Linkages between SAPs and the Environment," 25.

130. Mascarenhas, "Environment under Structural Adjustment," 42.

131. Sunseri, "Something Else to Burn," 631.

132. Kaale, "Trees for Village Forestry."

133. John Kimwaga, "The Vicious Circle of Deforestation," *Daily News*, February 7, 1985.

134. Kaale, "Trees for Village Forestry," iv–v.

135. Kunkle, "Contesting Globalization," 241.

136. Lee, *Making a World after Empire.*

137. Greg Grandin, "Down from the Mountain," *London Review of Books,* June 29, 2017, https://www.lrb.co.uk/v39/n13/greg-grandin-down-from-the-mountain.

138. Nyerere, "Unity for a New Order."

139. Kaufman, "Developing Countries Develop Self-Help."

140. Kaufman, "Developing Countries Develop Self-Help."

141. Amir Jamal, "Developing a Common Market," *Daily News,* December 23, 1985.

142. Hosier, "Economics of Deforestation in Eastern Africa," 129.

143. Hosier, "Economics of Deforestation in Eastern Africa," 129.

144. World Bank, *Tanzania: Issues and Options in the Energy Sector,* 68.

145. Tanzania Fiscal Year 1975–76 Budget, June 23, 1975 Telegram from Tanzania, Dar es Salaam to Department of state, Kenya Nairobi, Secretary of State accessed August 2018 https://wikileaks.org/plusd/cables/1975DARES02228_b.html

146. "Third World Urged to Cut Imports," *Daily News,* January 31, 1983.

147. "Daily News Reporter" and "Wrong Choices of Technology Deplored," *Daily News,* September 6, 1985.

148. From 1982 to 1983. Temu, Kaale, Maghembe, "Wood-Based Energy for Development in Tanzania," 4.

149. Nwgare, "Environmental Degradations and Fuelwood Consumption," 27.

150. Shechambo, "Urban Demand for Charcoal in Tanzania," 9.

151. Nindi, "State Intervention, Contradictions and Agricultural Stagnation in Tanzania," 131.

152. *Daily News,* June 15, 1979 as quoted in Nindi, "State Intervention, Contradictions and Agricultural Stagnation in Tanzania," 132.

153. Nindi, "State Intervention, Contradictions and Agricultural Stagnation in Tanzania," 132.

154. Nindi, "State Intervention, Contradictions and Agricultural Stagnation in Tanzania," 132.

155. Chachage, "Forms of Accumulation, Agriculture and Structural Adjustment in Tanzania," 222.

156. Tripp, *Changing the Rules,* 190.

157. Tripp, *Changing the Rules,* 33.

158. Mnzava, *Tree Planting in Tanzania,* 38.

159. It was also used in the expanding tobacco industry to cure leaves: 95 percent of Tanzanian tobacco was cured by wood and it took roughly 2.5 hectares of woodlands to make a hectare of tobacco. Openshaw and Food and Agriculture Organization of the United Nations, "Present Consumption and Future Requirements of Wood in Tanzania," 4.

160. Charcoal production and who has access to Tanzanian forests remains a fraught issue today. Sauli Giliard, "JPM Decries Indiscriminate Felling of Trees

for Charcoal," *Daily News*, July 24, 2017, http://dailynews.co.tz/index.php/home-news/51925-jpm-decries-indiscriminate-felling-of-trees-for-charcoal.

161. Some examples of this literature: Leach and Mearns, *Beyond the Woodfuel Crisis*. Arnold, and Persson, "Reassessing the Fuelwood Situation in Developing Countries," 379–83. Twine and Holdo, "Fuelwood Sustainability Revisited," 1766–76.

162. Johnsen, "Burning with Enthusiasm," 110.

163. Khazan, "Coal Burning in the U.S. and Europe Caused a Massive African Drought."

164. Charcoal generates about USD 650 million per year whereas coffee generates 60 million and tea 45 million. Peter and Sander, *Environmental Crisis or Sustainable Development Opportunity?*, vi.

Conclusion: Provisioning for an Unknown Future

1. Lofchie, "Agrarian Crisis and Economic Liberalisation in Tanzania," 451–475. Lofchie, *Political Economy of Tanzania*, 279–306. Boesen and Nordiska Afrikainstitutet, *Tanzania: Crisis and Struggle for Survival*. Hyden, *Beyond Ujamaa in Tanzania*.

2. "From the vantage point of the present, it is tempting to reduce the ujamaa experiment to a quixotic scheme and mere historical curiosity, at best, or to dismiss it as one of many examples of state authoritarianism confirming the generalized dysfunction of postcolonial African politics, at worst. Both perspectives define ujamaa not just by its ultimate failure to accomplish its intended goals but that it was inherently doomed to fail from its very inception." Lal, *African Socialism in Postcolonial Tanzania*, 4.

3. Goldstone and Obarrio, *African Futures*, 6. They are working here explicitly with the recent work of Roitman, *Anti-crisis*.

4. As Timothy Mitchell writes, a crisis narrative "simplifies changes in multiple fields, involving various agents, into a unique event, so that a single moment, with a single agent, appears responsible for a collapse of the old order." Mitchell, *Carbon Democracy*, 173.

5. For more on this see Brian Larkin, "The Form of Crisis and the Affect of Modernization," in Goldstone and Obarrio, *African Futures*, 39–50.

6. Mitchell, *Carbon Democracy*, 34.

7. Mitchell, *Carbon Democracy*, 34.

8. Coulson, *Tanzania: A Political Economy*, 274. For further discussion on foreign aid and expatriates see Aminzade, *Race, Nation and Citizenship in Post-colonial Africa*, 173–84.

9. Saul, *Revolutionary Traveller*, 33. Quoting from the document submitted by the "group of nine lecturers."

10. Saul, *Revolutionary Traveller*, 35.

11. Saul, *Revolutionary Traveller*, 36.

12. While the total rehaul of coursework was vetoed, there were some common core courses introduced to the curriculum to focus on general development problems.

13. Saul, *Revolutionary Traveller*, 60. To see what came of this in a compromise decision: Wield and Barker, "Course Bibliography," 385.

14. Of course, these are not always antithetical readings and many urban scholars have highlighted how reinvention of urban forms is a response to such failures.

15. Halberstam, *Queer Art of Failure*, 2–3.

16. White, *Railroaded*, 517.

17. Degani, "Shock Humor," 473–98.

BIBLIOGRAPHY

Archives and Libraries

Ardhi University Library, Dar es Salaam, Tanzania

British Institute in Eastern Africa, Nairobi, Kenya

British Library for Development Studies, University of Sussex, UK

British National Archives, Kew, UK

East Africana Library, Dar es Salaam, Tanzania

Herskovits Library, Northwestern University, Evanston, Illinois, USA

Institute for Development Studies Documentation Centre, University of Dar es Salaam, Tanzania

International Monetary Fund Archives, Washington, DC, USA

Ministry of Lands, Housing and Human Settlements Resource Centre, Dar es Salaam, Tanzania

Tanzanian National Archives, Dar es Salaam, Tanzania

Tanzanian National Library, Dar es Salaam, Tanzania

United States National Archives, College Park, Maryland, USA

Periodicals

Citizen (Tanzania)

Daily News (Tanzania)

Tanganyika Standard (Tanzania)

Nchi Yetu (Tanzania)

New York Times (United States)

New Yorker (United States)

Nationalist (Tanzania)

Sources

Abourahme, Nasser. "Assembling and Spilling-Over: Towards an 'Ethnography of Cement' in a Palestinian Refugee Camp." *International Journal of Urban and Regional Research* 39, no. 2 (2014): 200–17.

Agrawal, Arun. *Environmentality: Technologies of Government and the Making of Subjects.* Durham, NC: Duke University Press, 2005.

Aminzade, Ronald. *Race, Nation, and Citizenship in Post-colonial Africa: The Case of Tanzania*. New York: Cambridge University Press, 2013.

Anand, Nikhil. *Hydraulic City: Water and Infrastructures of Citizenship in Mumbai*. Durham, NC: Duke University Press, 2017.

Anand, Nikhil, Akhil Gupta, and Hannah Appel, eds. *The Promise of Infrastructure*. Durham, NC: Duke University Press, 2018.

Anderson, Benedict. *The Spectre of Comparisons: Nationalism, Southeast Asia and the World*. London: Verso, 1998.

Anderson, Warwick. "Crap on a Map, or Postcolonial Waste." *Postcolonial Studies* 13, no. 2 (2010): 169–78.

Andreasen, Manja Hoppe. "Population Growth and Spatial Expansion of Dar Es Salaam: An Analysis of the Rate and Spatial Distribution of Recent Population Growth in Dar Es Salaam." Working Paper 1, Rurban Africa: African Rural-City Connections. University of Copenhagen, 2013.

Armstrong, Allen. "Colonial and Neocolonial Urban Planning: Three Generations of Master Plans for Dar es Salaam Tanzania." *Utafiti* 8, no. 1 (1986): 43–66.

Arnold, M., and R. Persson. "Reassessing the Fuelwood Situation in Developing Countries." *International Forestry Review* 5, no. 4 (2003): 379–83.

Askew, Kelly. *Performing the Nation: Swahili Music and Cultural Politics in Tanzania*. Chicago: University of Chicago Press, 2002.

Askew, Kelly. "Sung and Unsung: Musical Reflections of Tanzanian Postcolonialisms." *Africa: Journal of International African Institute* 76, no. 1 (2006): 15–43.

Bagachwa, Mboya S. D. "Linkages between SAPs and the Environment: An Overview." In *Policy Reform and the Environment in Tanzania*, edited by Mboya S. D. Bagachwa and Festus Limbu, 17–36. Dar es Salaam: Dar es Salaam University Press, 1995.

Barnes, Teresa A. "The Fight for Control of African Women's Mobility in Colonial Zimbabwe, 1900–1939." *Signs* 17, no. 3 (1992): 586–608.

Barry, Andrew. *Material Politics: Disputes along the Pipeline*. Hoboken, NJ: Wiley, 2013.

Bates, Robert. *Markets and States in Tropical Africa: The Political Basis of Agricultural Policies*. Berkeley: University of California Press, 1981.

Benjamin, Walter. *Reflections: Essays, Aphorisms, Autobiographical Writing*. New York: Schocken Books, 1986.

Bertoncini-Zúbková, Elena. *Outline of Swahili Literature: Prose Fiction and Drama*. Leiden: Brill, 1989.

Bienefeld, M. A. "A Long-Term Housing Policy for Tanzania." Economic Research Bureau Paper 70.9. University College, Dar es Salaam.

Bienefeld, M. S., and R. H. Sabot. "The National Urban Mobility Employment and Income Survey of Tanzania." Ministry of Economic Affairs and Development Planning. Economic Research Bureau, University of Dar es Salaam, 1972.

Boesen, Jannik, and Nordiska Afrikainstitutet. *Tanzania: Crisis and Struggle for Survival*. Uppsala: Scandinavian Institute of African Studies, 1986.

Brennan, James. "Blood Enemies: Exploitation and Urban Citizenship in the Nationalist Political Thought of Tanzania, 1958–75." *Journal of African History* 47, no. 3 (2006): 389–413.

Brennan, James. *Taifa: Making Race and Nation in Urban Tanzania.* Athens: Ohio University Press, 2012.

Brennan, James, Andrew Burton, and Yusuf Lawi. *Dar es Salaam: Histories from an Emerging African Metropolis.* Dar es Salaam: Mkuki Na Nyota, 2007.

Briggs, John. "The Peri-urban Zone of Dar es Salaam, Tanzania: Recent Trends and Changes in Agricultural Land Use." *Transactions of the Institute of British Geographers* 16, no. 3 (1991): 319–31.

Briggs, John. "Villagisation and the 1974–6 Economic Crisis in Tanzania." *Journal of Modern African Studies* 17, no. 4 (1979): 695–702.

Briggs, John, and Davis Mwamfupe. "Peri-urban Development in an Era of Structural Adjustment in Africa: The City of Dar Es Salaam, Tanzania." *Urban Studies* 37, no. 4 (2000): 797–809.

Brownell, Emily. "Seeing Dirt in Dar es Salaam: Urban Citizenship and Waste in Postcolonial Tanzania." In *The Art of Citizenship in African Cities: Infrastructures and Spaces of Belonging,* edited by Mamadou Diouf and Rosalind Fredericks, 209–29. New York: Palgrave Macmillan, 2014.

Bryant, Raymond, and Sinéad Bailey. *Third World Political Ecology.* New York: Routledge, 1997.

Bryceson, Deborah. "A Century of Food Supply in Dar es Salaam: From Sumptuous Suppers for the Sultain to Maize Meal for a Million." In *Feeding African Cities: Studies in Regional Social History,* edited by Jane Guyer, 155–202. London: Routledge, 1987.

Bryceson, Deborah. *Food Insecurity and the Social Division of Labour in Tanzania, 1919–85.* New York: Palgrave Macmillan, 1990.

Burchell, Graham, Colin Gordon, and Peter Miller. *The Foucault Effect: Studies in Governmentality.* Chicago: University of Chicago Press, 1991.

Burke, Timothy. *Lifebuoy Men, Lux Women: Commodification, Consumption, and Cleanliness in Modern Zimbabwe.* Durham, NC: Duke University Press, 1996.

Burssens, Peter. "The (Non)Political Position of the Architecture of Anthony B. Almeida Between 1948 and 1975." In *ArchiAfrika Proceedings from the Conference on Modern Architecture in East Africa Around Independence, July 27–29, 2005, Dar es Salaam, Tanzania.* Utrecht: AchiAfrika, 2005. Accessed July 18, 2017. https://archnet.org/publications/4903.

Burton, Andrew. *African Underclass: Urbanisation, Crime and Colonial Order in Dar es Salaam.* Athens: Ohio University Press, 2005.

Burton, Andrew. "Haven of Peace Purged: Tackling the Undesirable and Unproductive Poor in Dar es Salaam, ca. 1950–1980s." *International Journal of African Historical Studies* 40, no. 1 (2007): 119–51.

Byerlee, Derek. "Rural-Urban Migration in Africa: Theory, Policy and Research Implications." *International Migration Review* 8, no. 4 (1974): 543–66.

Caldwell, Lynton K. "A World Policy for the Environment." *UNESCO Courier*, January 1973, 4–6, 32.

Callaci, Emily. "'Chief Village in a Nation of Villages': History, Race and Authority in Tanzania's Dodoma Plan." *Urban History* 43, no. 1 (2016): 96–116.

Callaci, Emily. *Street Archives and City Life: Popular Intellectuals in Postcolonial Tanzania*. Durham, NC: Duke University Press, 2017.

Callaci, Emily. "Street Textuality: Socialism, Masculinity, and Urban Belonging in Tanzania's Pulp Fiction Publishing Industry, 1975–1985." *Comparative Studies in Society and History* 59, no. 1 (2017): 183–210.

Capital Development Authority. *Building the National Capital: A Special Report to Mark the Fourth Anniversary of the Founding of the Chama Cha Mapinduzi*. Dodoma, Tanzania: Capital Development Authority, 1981.

Capital Development Authority. *Why Dodoma?* Pamphlet. Capital Development Authority, 1973.

Certeau, Michel de. *The Practice of Everyday Life*. Translated by Steven F. Rendall. Berkeley: University of California Press, 2011.

Chachage, C. S. L. "Forms of Accumulation, Agriculture and Structural Adjustment in Tanzania." In *Social Change and Economic Reform in Africa*, edited by Peter Gibbon, 215–43. Uppsala: Scandinavian Institute of African Studies, 1993.

Chaix, Jean Kim. "Is a Charcoal Crisis Looming for Tanzania?" *Charcoal Project*, January 19, 2010. http://www.charcoalproject.org/2010/01/is-a-woodfuel-and -charcoal-crisis-looming-for-tanzania/.

Chakrabarty, Dipesh. "The Muddle of Modernity." *American Historical Review* 116, no. 3 (2011): 663–75.

Chakrabarty, Dipesh. *Provincializing Europe: Postcolonial Thought and Historical Difference*. Princeton, NJ: Princeton University Press, 2007.

Chang, Jiat-Hwee. "Building a Colonial Technoscientific Network: Tropical Architecture, Building Science and the Politics of Decolonization." In *Third World Modernism: Architecture, Development and Identity*, edited by Duanfang Lu, 211–35. New York: Routledge, 2010.

Cliffe, Lionel, ed. *One-Party Democracy: The 1965 Tanzania General Election*. Nairobi: East African Publishing House, 1967.

Cloete, Stuart. "End of Era with Threat of Jungle Taking Over." *LIFE*, August 1, 1960.

Cohen, Shaul. *Planting Nature: Trees and the Manipulation of Environmental Stewardship in America*. Berkeley: University of California Press, 2004.

Cohen, Shaul. "The Politics of Planting: Israeli Palestinian Competition for Control of Land in the Jerusalem Periphery." Geography Research Paper No. 236. University of Chicago, 1993.

Collier, Stephen J. *Post-Soviet Social: Neoliberalism, Social Modernity, Biopolitics*. Princeton, NJ: Princeton University Press, 2011.

Collier, Stephen J., James Christopher Mizes, and Antina von Schnitzler, eds. "Public Infrastructures/Infrastructural Publics." Special issue, *Limn* 7 (July 2016).

Cooper, Frederick. "Africa and the World Economy." *African Studies Review* 24, no. 2–3 (1981): 1–86.

Cooper, Frederick. *Decolonization and African Society: The Labor Question in French and British Africa.* Cambridge: Cambridge University Press, 1996.

Cooper, Frederick. *On the African Waterfront: Urban Disorder and the Transformation of Work in Colonial Mombasa.* New Haven, CT: Yale University Press, 1987.

Cooper, Frederick, and Randall M. Packard. *International Development and the Social Sciences: Essays on the History and Politics of Knowledge.* Berkeley: University of California Press, 1997.

Coulson, Andrew. *Tanzania: A Political Economy.* 2nd ed. Oxford: Oxford University Press, 2013.

Cronon, William. *Nature's Metropolis: Chicago and the Great West.* New York: W.W. Norton, 1992.

Crowley, John. "The Politics of Belonging: Some Theoretical Considerations." In *The Politics of Belonging: Migrants and Minorities in Contemporary Europe*, edited by Andrew Geddes and Adrian Favell, 15–41. Aldershot: Ashgate, 1999.

Daley, Elizabeth. "Land and Social Change in a Tanzanian Village 1: Kinyanambo, 1920s–1990." *Journal of Agrarian Change* 5, no. 3 (2005): 363–404.

Dar es Salaam Ministry of Works. *Local Construction Industry Study: Summary Report.* Dar es Salaam: Ministry of Works, 1977.

Darkoh, M. B. K. "Desertification in Tanzania." *Geography* 67, no. 4 (1982): 320–31.

Davis, Diana K. "Deserts." In *Oxford Handbook for Environmental History*, edited by Andrew Isenberg, 108–32. New York: Oxford University Press, 2014.

Davis, Diana K. *Resurrecting the Granary of Rome: Environmental History and French Colonial Expansion in North Africa.* Athens: Ohio University Press, 2007.

Davis, Mike. *Planet of Slums.* London: Verso, 2005.

De Boek, Filip. "Spectral Kinshasa: Building the City through an Architecture of Words." In *Urban Theory Beyond the West: A World of Cities*, edited by Tim Edensor and Mark Jayne, 311–28. London: Routledge, 2012.

De Boeck, Filip, and Marie-Françoise Plissart. *Kinshasa: Tales of the Invisible City.* Leuven: Leuven University Press, 2014.

Decker, Alicia. "Idi Amin's Dirty War: Subversion, Sabotage and the Battle to Keep Uganda Clean, 1971–1979." *International Journal of African Historical Studies* 43, no. 3 (2010): 489–513.

Degani, Michael. "Shock Humor: Zaniness and the Freedom of Permanent Improvisation in Urban Tanzania." *Cultural Anthropology* 33, no. 3 (2018): 473–98.

Deleuze, Gilles, and Félix Guattari. *A Thousand Plateaus: Capitalism and Schizophrenia.* Translated by Brian Massumi. Minneapolis: University of Minnesota Press, 1987.

Derrida, Jacques. "La différance." *Bulletin de la Société Française de Philosophie* 62, no. 3 (1968): 73–101.

Diaz Olvera, Lourdes, Didier Plat, and Pascal Pochet. "Transportation and Access to Urban Services in Dar Es Salaam." Paper presented at CODATU Urban Mo-

bility for All conference, Lomé, Togo, November 12–15, 2002. https://halshs
.archives-ouvertes.fr/halshs-00088020/document.

Diof, Mamadou, and Rosalind Fredericks, eds. *The Arts of Citizenship in African Cities: Infrastructures and Spaces of Belonging.* New York: Palgrave, 2014.

Douglas, Mary. *Purity and Danger: An Analysis of Concepts of Pollution and Taboo.* London: Routledge, 2002.

Dove, Michael. "Forest Discourses in South and Southeast Asia: A Comparison with Global Discourses." In *Nature in the Global South: Environmental Projects in South and Southeast Asia,* edited by Paul Greenough and Anna Lowenhaupt Tsing, 103–23. Durham, NC: Duke University Press, 2003.

Eckholm, Erik. "The Other Energy Crisis: Firewood." Worldwatch Paper 1, Worldwatch Institute, September 1975.

Edwards, Paul N. *A Vast Machine: Computer Models, Climate Data, and the Politics of Global Warming.* Cambridge, MA: MIT Press, 2010.

Ehrlich, Paul. *The Population Bomb.* New York: Ballantine Books, 1976.

Enevoldsen, Thyge, Helle Fischer, Niels Fold, and Steen Folke. *India's Export of Capital Goods to Tanzania—In a Development Context.* Copenhagen: University of Copenhagen Institute for Geography, 1988.

Englund, Harri. "The Village in the City, The City in the Village: Migrants in Lilongwe." *Journal of Southern African Studies* 28, no. 1 (2002): 137–54.

Enwezor, Okwui, Carlos Basualdo, Ute Meta Bauer, Susanne Ghez, Sarat Maharaj, Mark Nash, and Octavio Zaya, eds. *Under Siege, Four African Cities, Freetown, Johannesburg, Kinshasa, Lagos.* Documenta 11_Platform 4. Ostfildern-Ruit: Hatje Cantz, 2002.

Ergas, Zaki. "Why Did the Ujamaa Village Policy Fail? Towards a Global Analysis." *Journal of Modern African Studies* 18, no. 3 (1980): 387–410.

Eriken, Svein, and Bo Ronnstedt. "Building Cost Indices." Working Reports No. 28, Building Research Unit, Dar es Salaam, 1982.

Esty, Joshua D. "Excremental Postcolonialism." *Contemporary Literature* 40, no. 1 (1999): 22–59.

Evans, Brad, and Julian Reid. *Resilient Life: The Art of Living Dangerously.* New York: Polity, 2014.

Fair, Laura. "Drive-In Socialism: Debating Modernities and Development in Dar es Salaam, Tanzania." *American Historical Review* 118, no. 4 (2013): 1077–104.

Fair, Laura. *Reel Pleasures: Cinema Audiences and Entrepreneurs in Twentieth-Century Urban Tanzania.* Athens: Ohio University Press, 2017.

Fanon, Frantz. *Wretched of the Earth.* Translated by Constance Farrington. New York: Grove Press, 1963.

Farvar, M. Taghi, and John P. Milton, eds. *The Careless Technology: Ecology and International Development.* Garden City, NY: Natural History Press, 1972.

Ferguson, James. *The Anti-politics Machine: "Development," Depoliticization, and Bureaucratic Power in Lesotho.* Cambridge: Cambridge University Press, 1990.

Ferguson, James. "The Country and the City on the Copperbelt." *Cultural Anthropology* 7, no. 1 (1992): 80–92.

Ferguson, James. *Expectations of Modernity: Myths and Meanings of Urban Life on the Zambian Copperbelt*. Berkeley: University of California Press, 1999.

Foley, Gerald, Patricia Moss, and Lloyd Timberlake. *Stoves and Trees: How Much Wood Would a Woodstove Save If a Woodstove Could Save Wood?* Berkeley, CA: International Institute for Environment and Development and Earthscan, 1984.

Folke, Steen, Niels Fold, and Thyge Enevoldsen. *South-South Trade and Development: Manufactures in the New International Division of Labour*. New York: St. Martin's, 2016.

Food and Agriculture Organization and United Nations Development Programme. *Survey of the Kisutu Market, Dar es Salaam*. Project SF TAN 27 Ministry of Agriculture, Marketing Development Bureau FAO-UNDP, August 1972.

Food and Agriculture Organization and United Nations Environment Programme. *Tropical Forest Resources Assessment Project: Forest Resources of Tropical Asia*. UN 32/6.1301-78-04 Technical Report 3. Rome: Food and Agriculture Organization and United Nations Environment Programme, 1981. http://www.fao.org/3/ad908e/ad908e00.htm.

The Forest Ordinance of Tanzania (1957): Principal Legislation—Cap 389 of 1957. Accessed May 9, 2017. http://www.tfs.go.tz/uploads/Forest_Ordinance_1957.pdf.

Forty, Adrian. *Concrete and Culture: A Material History*. London: Reaktion Books, 2012.

Foster, John Bellamy. *Marx's Ecology: Materialism and Nature*. New York: Monthly Review, 2000.

Fredericks, Rosalind. *Garbage Citizenship: Vibrant Infrastructures of Labor in Dakar, Senegal*. Durham, NC: Duke University Press, 2018.

Freeman, Donald B. "Survival Strategy or Business Training Ground? The Significance of Urban Agriculture for the Advancement of Women in African Cities." *African Studies Review* 36, no. 3 (1993): 1–22.

French, Daniel. *When They Hid the Fire: A History of Electricity and Invisible Energy in America*. Pittsburgh: University of Pittsburgh Press, 2017.

Freyhold, Michaela von. "Notes on Tanzanian Industrial Workers." *Tanzania Notes and Records* 81 (1977): 15–22.

Freyhold, Michaela von. *Ujamaa Villages in Tanzania: Analysis of a Social Experiment*. London: Heinemann, 1979.

Friedman, Andrew. "The Global Postcolonial Moment and the American New Town: India, Reston, Dodoma." *Journal of Urban History* 38, no. 3 (2012): 553–76.

Gandhi, Indira. "Indira Gandhi's Speech at the Stockholm Conference in 1972: Man and Environment." Speech delivered at the plenary session of the United Nations Conference on Human Environment, Stockholm, June 14, 1972. *LASU-LAWS*

Environmental Blog, July 18, 2012. http://lasulawsenvironmental.blogspot
.de/2012/07/indira-gandhis-speech-at-stockholm.html.

Gandy, Matthew. *Concrete and Clay: Reworking Nature in New York City.* Cambridge,
MA: MIT Press, 2003.

Gandy, Matthew. *The Fabric of Space: Water, Modernity, and the Urban Imagination.*
Cambridge, MA: MIT Press, 2014.

Gandy, Matthew. "Vicissitudes of Nature: Transitions and Transformations at a
Global Scale." *Radical History Review* 2010, no. 107 (2010): 178–84.

Geiger, Susan. *TANU Women: Gender and Culture in the Making of Tanganyikan Nationalism, 1955–1965.* Portsmouth, NH: Heinemann, 1997.

George, Abosede. *Making Modern Girls: A History of Girlhood, Labor, and Social Development in Colonial Lagos.* Athens: Ohio University Press, 2014.

Geschiere, Peter, and Josef Gugler. "The Urban-Rural Connection: Changing Issues
of Belonging and Identification." *Africa* 68, no. 3 (1998): 309–19.

Gille, Zsuzsa. *From the Cult of Waste to the Trash Heap of History: The Politics of Waste
in Socialist and Postsocialist Hungary.* Bloomington: Indiana University Press,
2007.

Gissen, David. *Subnature: Architecture's Other Environments.* New York: Princeton
Architectural Press, 2012.

Goldstone, Brian, and Juan Obarrio. *African Futures: Essays on Crisis, Emergence, and
Possibility.* Chicago: University of Chicago, 2017.

Grace, Joshua Ryan. "Modernization *Bubu:* Cars, Roads, and the Politics of Development in Tanzania, 1870s–1980s." PhD diss., University of Michigan, 2013.

Graham, Stephen, and Nigel Thrift. "Out of Order: Understanding Repair and Maintenance." *Theory, Culture & Society* 24, no. 3 (2007): 1–25.

Grandin, Greg. "Down from the Mountain." *London Review of Books,* June 29, 2017.

Gruen, Victor. *The Heart of Our Cities: The Urban Crisis: Diagnosis and Cure.* New
York: Simon and Schuster, 1964.

Gutschow, Kai K. "*Das Neue Afrika*: Ernst May's 1947 Kampala Plan as Cultural
Programme." In *Colonial Architecture and Urbanism in Africa: Intertwined and
Contested Histories,* edited by Fassil Demissie, 373–406. London: Ashgate, 2009.

Halberstam, Jack (see also Judith Halberstam). *The Queer Art of Failure.* Durham,
NC: Duke University Press, 2011.

Hanlon, Joseph. "Where Concrete Meets Cane." *New Scientist,* August 31, 1978,
627–29.

Hart, Jennifer. *Ghana on the Go: African Mobility in the Age of Motor Transportation.*
Bloomington: Indiana University Press, 2016.

Hart, Keith. "Migration and Tribal Identity among the Frafras of Ghana." *Journal of
Asian and African Studies* 6, no. 1 (1971): 21–36.

Harvey, David. *Social Justice in the City.* Athens: University of Georgia Press, 2009.

Harvey, Penny, and Hannah Knox. "Abstraction, Materiality and the 'Science of the
Concrete' in Engineering Practice." In *Material Powers: Cultural Studies, History*

and the Material Turn, edited by Tony Bennett and Patrick Joyce, 124–41. New York: Routledge, 2010.

Havenvik, Kjell, Rune Skarstein, and Smauel Wangwe. *Small Scale Industrial Sector Survey—Tanzania: Review of Experiences and Recommendations for the Future*. Report presented to the Ministry of Industries and Trade, 1985. East Africana Library, University of Dar es Salaam.

Hazlewood, Arthur. *Economic Integration: The East African Experience*. New York: St. Martin's, 1975.

Herrera, Veronica. *Water and Politics: Clientalism and Reform in Urban Mexico*. Ann Arbor: University of Michigan Press, 2017.

Herrmann, S. M., and C.F. Hutchinson. "The Changing Contexts of the Desertification Debate." *Journal of Arid Environments* 63, no. 3 (2005): 538–55.

Heynen, Nik, Maria Kaika, and Erik Swyngedouw, eds. *In the Nature of Cities: Urban Political Ecology and the Politics of Urban Metabolism*. New York: Routledge, 2006.

Higgs, Kerryn. *Collision Course: Endless Growth on a Finite Planet*. Cambridge, MA: MIT Press, 2014.

Hoag, Heather J. *Developing the Rivers of East and West Africa: An Environmental History*. New York: Bloomsbury, 2013.

Holston, James. *The Modernist City: An Anthropological Critique of Brasilia*. Chicago: University of Chicago Press, 1989.

Hosier, Richard H. "The Economics of Deforestation in Eastern Africa." *Economic Geography* 64, no. 2 (1988): 121–36.

Hoyle, B. S. "African Politics and Port Expansion at Dar es Salaam." *Geographical Review* 68, no. 1 (1978): 31–50.

Hoyle, B. S. "African Socialism and Urban Development: The Relocation of the Tanzanian Capital." *Tijdschrift Voor Economische en Sociale Geografie* 70, no. 4 (1979): 207–16.

Huba, Nguluma. "Access to Housing through Cooperatives: Potentials and Challenges from Tanzania." *International Journal of Humanities, Social Sciences and Education* 3, no. 9 (2016): 100–114.

Hull, Matthew. "Communities of Place, Not Kind: American Technologies of Neighborhood in Postcolonial Delhi." *Comparative Studies in Society and History* 53, no. 4 (2011): 757–90.

Humphrey, Caroline. "Ideology in Infrastructure: Architecture and Soviet Imagination." *Journal of the Royal Anthropological Institute* 11, no. 1 (2005): 39–58.

Hunt, Kathleen P. "'It's More Than Planting Trees, It's Planting Ideas': Ecofeminist Praxis in the Green Belt Movement." *Southern Communication Journal* 79, no. 3 (2014): 235–49.

Hunter, Emma. *Political Thought and the Public Sphere in Tanzania: Freedom, Democracy and Citizenship in the Era of Decolonization*. Cambridge: Cambridge University Press, 2015.

Hunter, Emma. "Voluntarism, Virtuous Citizenship, and Nation-Building in Late Colonial and Early Postcolonial Tanzania." *African Studies Review* 58, no. 2 (2015): 43–61.

Hurst, Andrew. "State Forestry and Spatial Scale in the Development Discourses of Post-Colonial Tanzania: 1961–1971." *Geographical Journal* 169, no. 4 (2003): 358–69.

Hutton, John. *Urban Challenge in East Africa.* Nairobi: East African Publishing House, 1970.

Hyden, Göran. *Beyond Ujamaa in Tanzania: Underdevelopment and an Uncaptured Peasantry.* London: Heinemann, 1980.

Iliffe, John. *The African Poor: A History.* Cambridge: Cambridge University Press, 1987.

Immerwahr, Daniel. *Thinking Small: The United Sates and the Lure of Community Development.* Cambridge, MA: Harvard University Press, 2015.

Ince, Margaret, ed. *Water and Sanitation in Africa.* Proceedings of the 11th Water and Engineering for Developing Countries Conference, Dar es Salaam, April 15–19, 1985. https://www.ircwash.org/sites/default/files/71-WEDC85-6424.pdf.

Ingold, Tim. *The Perception of the Environment.* New York: Routledge, 2000.

Ivaska, Andrew. *Cultured States: Youth Gender, and Modern Style in 1960s Dar es Salaam.* Durham, NC: Duke University Press, 2011.

Jackson, Steven J. "Rethinking Repair." In *Media Technologies: Essays on Communication, Materiality and Society*, edited by Tarleton Gillespie, Pablo J. Boczkowski, and Kirsten A. Foot, 221–40. Cambridge, MA: MIT Press, 2014.

Jean-Baptiste, Rachel. *Conjugal Rights: Marriage, Sexuality, and Urban Life in Colonial Libreville, Gabon.* Athens: Ohio University Press, 2014.

Jeffrey, Craig. *Timepass: Youth, Class, and the Politics of Waiting in India.* Stanford, CA: Stanford University Press, 2010.

Jennings, Michael. "'Development is Very Political in Tanzania': Oxfam and the Chunya Integrated Development Programme 1972–96." In *Readings in the International Relations of Africa*, edited by Tom Young, 77–92. Bloomington: Indiana University Press, 2016.

Jennings, Michael. *Surrogates of the State: NGOs, Development, and Ujamaa in Tanzania.* Bloomfield, CT: Kumarian, 2007.

Johnsen, Fred Håkon. "Burning with Enthusiasm: Fuelwood Scarcity in Tanzania in Terms of Severity, Impacts and Remedies." *Forum for Development Studies* 26, no. 1 (1999): 107–31.

Jones, Frank E. "Operation and Maintenance Services for Dar es Salaam Sewerage and Sanitation." In *Water and Sanitation in Africa*, proceedings of the 11th Water and Engineering for Developing Countries Conference, Dar es Salaam, April 15–19, 1985, edited by Margaret Ince, 19–22. https://www.ircwash.org/sites/default/files/71-WEDC85-6424.pdf.

Kaale, B. K. "Trees for Village Forestry." Dar es Salaam: Forest Division, Ministry of Lands, Natural Resources and Tourism, 1984.

Kaitilla, Sababu. "The Influence of Environmental Variables on Building Material Choice: The Role of Low-Cost Building Materials on Housing Improvement." *Revue Architecture et Comportement/Architecture and Behaviour* 7, no. 3 (1991): 205–22.

Kammen, Dan. "Research, Development and Commercialization of the Kenya Ceramic Jiko and other Improved Biomass Stoves in Africa." *Improved Biomass Cooking Stoves.* http://stoves.bioenergylists.org/content/research-development-and-commercialization-kenya-ceramic-jiko-and-other-improved-biomass-sto.

Kanyama, Ahmad A. "Can the Urban Housing Problem be Solved Through Physical Planning? An Analysis Based on Experience from Dodoma, Tanzania." PhD diss., Tekniska Høgskolan, Sweden, 1995.

Kaplan, Robert D. "The Coming Anarchy." *Atlantic Monthly*, February 1994.

Kaplinsky, Raphael. *The Economies of the Small.* Washington, DC: Intermediate Technology and Appropriate Technology International, 1990.

Katz, Cindi. "On the Grounds of Globalization: A Topography for Feminist Political Engagement." *Signs* 26, no. 4 (2001): 1213–34.

Kaufman, Michael T. "Developing Countries Develop Self-Help." *New York Times*, February 28, 1982, sec. A, New York late edition.

Kent, David W., and Paul S. D. Mushi. "The Education and Training of Artisans for the Informal Sector in Tanzania." Education Research Paper No. 18, Education Division of the Overseas Development Administration, London, 1996. http://webarchive.nationalarchives.gov.uk/20090225031945/http://www.dfid.gov.uk/pubs/files/edtrainartedpaper18.pdf.

Kent, Susan. *Domestic Architecture and the Use of Space: An Interdisciplinary Cross-Cultural Study.* Cambridge: Cambridge University Press, 1993.

Kenya Woodfuel Development Programme. *So, Firewood can Wreck a Home?* Ministry of Energy and Regional Development, 1985.

Kerner, Donna O. "'Hard Work' and Informal Sector Trade in Tanzania." In *Traders versus the State: Anthropological Approaches to Unofficial Economies*, edited by Gracia Clark, 41–56. Boulder, CO: Westview, 1988.

Kessy, Joseph D. "Promoting Good Urban Governance at the Community Level." In *Urbanising Tanzania: Issues, Initiatives and Priorities*, edited by Suleiman Ngware and J. M. Lusugga Kironde, 133–36. Dar es Salaam: Dar es Salaam University Press, 2000.

Khazan, Olga. "Coal Burning in the U.S. and Europe Caused a Massive African Drought." *Atlantic Monthly*, June 2013.

Kirobo, A. Y. *Film-House in Manzese, Dar es Salaam: practical experiences from the construction of a demonstration house for the film 'Jijengee Nyumba Bora' (Build a Better House).* Working Reports No. 13, Building Research Unit, University of Dar es Salaam, 1977.

Kironde, J. M. Lusugga. "The Evolution of Land Use Structure of Dar es Salaam: A Study in the Effects of Land Policy." PhD diss., University of Nairobi, 1994.

Kironde, J. M. Lusugga. "Race, Class and Housing in Dar es Salaam: The Colonial Impact of Land Use Structure 1891–1961." In *Dar es Salaam: Histories from an Emerging Metropolis*, edited by James R. Brennan, Andrew Burton and Yusuf Lawi, 97–117. Dar es Salaam: Mkuki na Nyota, 2007.

Kironde, J. M. Lusugga. "Rent Control Legislation and the National Housing Corporation in Tanzania, 1985–1990." *Canadian Journal of African Studies / Revue Canadienne des Études Africaines* 26, no. 2 (1992): 306–27.

Kneitz, Agnes, ed. "The Country and the City." Special issue, *Global Environment* 9, no. 1 (2016).

Koenigsberger, Otto H. "New Towns in India." *Town Planning Review* 23, no. 2 (1952): 94–132.

Kombe, Jackson W. M. "The Demise of Public Urban Land Management and the Emergence of Informal Land Markets in Tanzania: A Case of Dar-es-Salaam City." *Habitat International* 18, no. 1 (1994): 23–43.

Kombe, Wilbard Jackson. "Land Use Dynamics in Peri-urban Areas and Their Implications on the Urban Growth and Form: The Case of Dar es Salaam, Tanzania." *Habitat International* 29, no. 1 (2005): 113–35.

Kombe, Wilbard J., and Volker Kriebich. "Informal Land Management in Tanzania and the Misconception about its Illegality." Paper presented at the ESF/N-Aerus Annual Workshop "Coping with Informality and Illegality in Human Settlements in Developing Countries," Leuven and Brussels, May 23–26, 2001.

Konde, Hadji S. *Press Freedom in Tanzania*. Arusha, Tanzania: Eastern Africa, 1984.

Konings, Piet, and D. Foeken. *Crisis and Creativity: Exploring the Wealth of the African Neighbourhood*. Leiden: Brill, 2006.

Kulaba, S. M. *Housing, Socialism and National Development in Tanzania: A Policy Framework*. Occasional paper, Bouwcentrum International Education and CHS Tanzania, Rotterdam, 1981.

Kunkle, Sönke. "Contesting Globalization: The United Nations Conference on Trade and Development and the Transnationalization of Sovereignty." In *International Organizations and Development, 1945–1990*, edited by Marc Frey, Sönke Kunkel, and Corinna R. Unger, 240–58. London: Palgrave Macmillan, 2014.

Kuuya, Masette P. *Import Substitution as an Industrial Strategy: The Tanzania Case*. Dar es Salaam: Economic Research Bureau, University of Dar es Salaam, 1976.

Lagos/Koolhaas. Directed by Bregtje an der Haak. Brooklyn: Icarus Films, 2002.

Lal, Priya. *African Socialism in Postcolonial Tanzania: Between the Village and the World*. Cambridge: Cambridge University Press, 2015.

Lal, Priya. "Militants, Mothers, and the National Family: 'Ujamaa,' Gender, and Rural Development in Postcolonial Tanzania." *Journal of African History* 51, no. 1 (2010): 1–20.

Lall, Sanjaya. *Multinationals, Technology and Exports: Selected Papers*. New York: Palgrave, 1985.

Larkin, Brian. *Signal and Noise: Media, Infrastructure, and Urban Culture in Nigeria.* Durham, NC: Duke University Press, 2008.

Latham, Michael E. *The Right Kind of Revolution: Modernization, Development, and U.S. Foreign Policy from the Cold War to the Present.* Ithaca, NY: Cornell University Press, 2010.

Leach, Gerald L., and Robin Mearns. *Beyond the Woodfuel Crisis: People, Land and Trees in Africa.* London: Earthscan, 1988.

Lee, Christopher, ed. *Making a World after Empire: The Bandung Moment and Its Political Afterlives.* Athens: Ohio University Press, 2010.

Lee-Smith, Diana, and Catalina Hinchey Trujillo. "The Struggle to Legitimize Subsistence: Women and Sustainable Development." *Environment and Urbanization* 4, no. 1 (1992): 77–84.

Leslie, J. A. K. *A Survey of Dar es Salaam.* Oxford: Oxford University Press, 1963.

Levine, Katherine. "The TANU Ten-House Cell System." In *Socialism in Tanzania: An Interdisciplinary Reader,* edited by Lionel Cliffe and John Saul, 328–37. Nairobi: East African Publishing House, 1972.

Li, Tania Murray. "Governmentality." *Anthropologica* 49, no. 2 (2007): 275–81.

Little, Richard G. "Controlling Cascading Failure: Understanding the Vulnerabilities of Interconnected Infrastructures." *Journal of Urban Technology* 9, no. 1 (2002): 109–23.

Livengood, Robert Mark. "Mafundi Chuma and Folk Recycling in Dar es Salaam: Case Studies of Material Behavior in Urban Tanzania." PhD diss., University of California, Los Angeles, 2001.

Lockrem, Jessica, and Adonio Lugo. "Infrastructure." *Cultural Anthropology.* Accessed September 17, 2017. https://culanth.org/curated_collections/11-infrastructure.

Lofchie, Michael F. "Agrarian Crisis and Economic Liberalisation in Tanzania." *Journal of Modern African Studies* 16, no. 3 (1978): 451–75.

Lofchie, Michael. "Political and Economic Origins of African Hunger." *Journal of Modern African Studies* 13, no. 4 (1975): 551–67.

Lofchie, Michael. *The Political Economy of Tanzania: Decline and Recovery.* Philadelphia: University of Pennsylvania Press, 2014.

Lonsdale, John. "The Moral Economy of Mau Mau." In *Unhappy Valley: Conflict in Kenya and Africa,* edited by Bruce Berman and John Lonsdale, 265–504. London: James Currey, 1992.

Low, D. A., and John M. Lonsdale. "Introduction: Towards the New Order, 1945–1963." In *History of East Africa,* vol. 3, edited by D. A. Low and Alison Smith, 1–63. Oxford: Clarendon, 1976.

Loxley, John, and John S. Saul. "Multinationals, Workers and the Parastatals in Tanzania." *Review of African Political Economy* no. 2 (1975): 54–88.

Lupala, Aldo, and John Lupala. "The Conflict between Attempts to Green Arid Cities and Urban Livelihoods: The Case of Dodoma, Tanzania." *Journal of Political Ecology* 10, no. 1 (2003): 25–35.

Lynch, Kenneth. "Urban Fruit and Vegetable Supply in Dar es Salaam." *Geographical Journal* 160, no. 3 (1994): 307–18.

Mabogunje, Akin. "Urban Planning and the Post-colonial State in Africa: An Overview." *African Studies Review* 33, no. 2 (1990): 121–203.

Macekura, Stephen. *Of Limits and Growth: The Rise of Global Sustainable Development in the Twentieth Century*. Cambridge: Cambridge University Press, 2015.

Maddox, Gregory. "Leave Wagogo, You Have No Food: Famine and Survival in Ugogo, Tanzania, 1916–1961." PhD diss., Northwestern University, 1988.

Maliyamkono, T. L., and Mboya S. D. Bagachwa. *The Second Economy in Tanzania*. London: James Currey, 1990.

Malkki, Liisa. "National Geographic: The Rooting of Peoples and the Territorialization of National Identity among Scholars and Refugees." *Cultural Anthropology* 7, no. 1 (1992): 24–44.

Many Words for Modern: A Survey of Modern Architecture in Tanzania. Directed by Jord den Hollander. Netherlands: ArchiAfrika and Jord den Hollander Film, 2007.

Marshall, Macklin and Monaghan, Ltd. *Dar es Salaam Master Plan*. Don Mills, Ontario: Marshall Macklin Monoghan Ltd, 1979.

Marshall, Macklin and Monaghan, Ltd. *Dar es Salaam Master Plan Summary*. Don Mills, Ontario: Marshall Macklin Monoghan Ltd, 1979.

Marshall, Macklin and Monaghan, Ltd. *Dar es Salaam Master Plan Technical Supplement Sewage Collection and Disposal*. Don Mills, Ontario: Marshall, Macklin Monoghan, 1979.

Martin, Phyllis M. "Contesting Clothes in Colonial Brazzaville." *Journal of African History* 35, no. 3 (1994): 401–26.

Martinez-Alier, Joan. "Ecology and the Poor: A Neglected Dimension of Latin American History." *Journal of Latin American Studies* 23, no. 3 (1991): 621–39.

Mascarenhas, Adolfo. "The Environment under Structural Adjustment with Specific Reference to Semi-Arid Areas." In *Policy Reform and the Environment in Tanzania*, edited by Mboya S. D. Bagachwa, Festus Limbu, Friedrich-Ebert-Stiftung, and National Economic Policy Workshop, 37–56. Dar es Salaam, Tanzania: Dar es Salaam University Press, 1995.

Masembejo, Lameki M., and Nkya J. W. Tumsiph. *Upgrading in Dar es Salaam*. Gavle, Sweden: National Swedish Institute for Building Research, 1981.

Mavhunga, Clapperton Chakanetsa. "*Cidades Esfumaçadas*: Energy and the Rural-Urban Connection in Mozambique." *Public Culture* 25, no. 2 (2013): 261–71.

Max, John A. O. *The Development of Local Government in Tanzania*. Dar es Salaam: Educational Publishers and Distributors, 1991.

Mayer, Albert. *Pilot Project, India: The Story of Rural Development at Etawah, Uttar Pradesh*. Westport, CT: Greenwood, 1973.

Mazrui, Ali K. "Tanzaphilia: A Diagnosis." *Transition* 31, no. 6 (1967): 20–26.

Mbilinyi, Marjorie. "'City' and 'Countryside' in Colonial Tanganyika." *Economic and Political Weekly* 20, no. 43 (1985): WS88–96.

Mbilinyi, S. M., and A. C. Mascarenhas. *The Sources and Marketing of Cooking Bananas in Tanzania.* Paper 69.14, Economic Research Bureau, Dar es Salaam, 1969.

McAuslan, J. P. W. B. "Law and Lawyers in Urban Development: Some Reflections from Practice." *Third World Legal Studies* 1 (1982): article 5.

McAuslan, Patrick. *The Ideologies of Planning Law.* Oxford: Pergamon, 1980.

McCann, James C. "The Plow and the Forest: Narratives of Deforestation in Ethiopia, 1840–1992." *Environmental History* 2, no. 2 (1997): 138–59.

McLees, Leslie. "Access to Land for Urban Farming in Dar Es Salaam, Tanzania: Histories, Benefits and Insecure Tenure." *Journal of Modern African Studies* 49, no. 4 (2011): 601–24.

McNeill, J. R., and Peter Engelke. *The Great Acceleration: An Environmental History of the Anthropocene since 1945.* Cambridge, MA: Belknap Press of Harvard University Press, 2016.

McNeur, Catherine. *Taming Manhattan: Environmental Battles in the Antebellum City.* Cambridge, MA: Harvard University Press, 2014.

Meadows, Donella, Dennis L. Meadows, Jørgen Randers, and William W. Behrens. *Limits to Growth: A Report for the Club of Rome's Project on the Predicament of Mankind.* New York: Signet, 1972.

Mercer, Claire. "Landscapes of Extended Ruralisation: Postcolonial Suburbs in Dar es Salaam, Tanzania." *Transactions of the Institute of British Geographers* 42, no. 1 (2017): 72–83.

Mesaki, Simeon. "Operation Pwani: Kisarawe District—Implementation, Problems, Prospects." MA thesis, University of Dar es Salaam, 1979.

Mesaki, Simeon, and August Nimtz. "The Politics of Capital Relocation: Dodoma, Tanzania." Paper presented at the annual meeting of the African Studies Association, Denver, Colorado, 1987.

Mihyo, Njalai Zukiva. "Women, Work and Struggle: A Case Study of Tanzania Friendship (Urafiki) Textile Mill." MA thesis, Institute of Social Studies, The Hague.

Mihyo, Paschal. "The Struggle for Workers' Control in Tanzania." *Review of African Political Economy* no. 4 (1975): 62–84.

Mitchell, James C. *The Kalela Dance: Aspects of Social Relationships among Urban Africans in Northern Rhodesia.* Manchester: Manchester University Press, 1956.

Mitchell, Timothy. *Carbon Democracy: Political Power in the Age of Oil.* New York: Verso, 2011.

Mkalawa, Charles Cosmas, and Pan Haixiao. "Dar es Salaam City Temporal Growth and Its Influence on Transportation." *Urban, Planning and Transport Research* 2, no. 1 (2014): 423–46.

Mnzava, E. M. *Tree Planting in Tanzania: A Voice from Villagers, Forest Division Dar es Salaam, 1983.* Stockholm: Swedish International Development Authority, 1984.

Mnzava, E. M. "Village Industries vs. Savanna Forests." FAO Paper at the United Nations Conference on New and Renewable Sources of Energy, Nairobi, August 1981.

Molohan, M. J. B. *Detribalization: A Study of the Areas of Tanganyika where Detribalized Persons are Living, with Recommendations as to the Administrative and Other Measures Required to Meet the Problems Arising Therein.* Dar es Salaam: Government Printer, 1959.

Molony, Thomas. *Nyerere: The Early Years.* London: James Currey, 2014.

Monson, Jamie. *Africa's Freedom Railway: How a Chinese Development Project Changed Lives and Livelihoods in Tanzania.* Bloomington: Indiana University Press, 2009.

Monson, Jamie. "*Maisha*: Life History and the History of Livelihood Along the TAZARA Railway in Tanzania." In *Sources and Methods in African History: Spoken, Written, Unearthed,* edited by Toyin Falola and Christian Jennings, 312–29. Rochester: University of Rochester Press, 2004.

Moorman, Marissa J. *Intonations: A Social History of Music and Nation in Luanda, Angola, from 1945 to Recent Times.* Athens: Ohio University Press, 2008.

Morton, David. *Age of Concrete: Housing and the Shape of Aspiration in the Capital of Mozambique.* Athens: Ohio University Press, 2019.

Morton, David. "Chamanculo in Reeds, Wood, Zinc and Concrete." *SLUM Lab* 9 (2014): 43–46.

Mtei, Edwin. *From Goatherd to Governor: The Autobiography of Edwin Mtei.* Dar es Salaam: Mkuki na Nyota, 2009.

Mushi, S. S. "Regional Development Through Rural Urban Linkages: The Dar-es Salaam Impact Region." PhD diss., University of Dortmund, 2003.

Mutasingwa, D. R. "The History of Dar es Salaam: The Study of Magomeni." PhD diss., University of Dar es Salaam, 1978/1979.

Mwakikagile, Godfrey. *Nyerere and Africa: End of an Era.* Pretoria, South Africa: New Africa, 2007.

Mwakysusa, A.R. "The Penetration of TANU into Industries: A Case Study on Friendship Textile Mill, TANU Branch." PhD diss., University College of Dar es Salaam, 1970.

Mwandosya, M. J., and M. L. P. Luhanga. "Energy Use Patterns in Tanzania." *Ambio* 14, no. 4/5 (1985): 237–41.

Myers, Garth. *African Cities: Alternative Visions of Urban Theory and Practice.* New York: Zed Books, 2011.

Nagendra, Harini. *Nature in the City: Bengaluru in the Past, Present and Future.* New Delhi: Oxford University Press, 2016.

Nash, Robert, and Cecilia Luttrell. "Crisis to Context: The Fuelwood Debate." Grey Literature, Overseas Development Institute Forest Policy and Environment Programme, March 2006. https://www.odi.org/sites/odi.org.uk/files/odi-assets/publications-opinion-files/3773.pdf.

National Research Council. *Firewood Crops: Shrub and Tree Species for Energy Production.* Washington, DC: The National Academies Press, 1980. https://doi.org/10.17226/21317.

Ndatulu, [Mrs.], and H. B. Makileo. *Housing Cooperatives in Tanzania.* Dar es Salaam: Building Research Unit, 1989.

Neumann, Roderick P. "Forest Rights, Privileges and Prohibitions: Contextualising State Forestry Policy in Colonial Tanganyika." *Environment and History* 3, no. 1 (1997): 45–68.

Nindi, B. C. "State Intervention, Contradictions and Agricultural Stagnation in Tanzania: Cashew Nut vs Charcoal Production." *Public Administration and Development* 11, no. 2 (1991): 127–34.

Nkonoki, Simon R. *Energy Crisis of the Poor in Tanzania: A Research Report of the Tanzania Rural Energy Consumption.* New York: Rockefeller Foundation, 1983.

Nnkya, Anna. "Housing and Design in Tanzania." PhD diss., Royal Danish Academy of Art, School of Architecture, 1984.

Nnkya, Tumsifu. *Shelter Co-operatives in Tanzania: Contributions of the Co-operative Sector to Shelter Development.* Nairobi: United Nations Centre for Human Settlements and International Co-operative Alliance, 2001.

Nuttall, Sarah, and Achille Mbembe. *Johannesburg: The Elusive Metropolis.* Durham, NC: Duke University Press, 2008.

Nwgare, Neema. "Environmental Degradations and Fuelwood Consumption: Its Impact on the Lives of Women and Their Families." PhD diss., University of Minnesota, 1996.

Nyerere, Julius K. *The Arusha Declaration and TANU's Policy on Socialism and Self-Reliance.* Dar es Salaam: TANU, 1967.

Nyerere, Julius. "The Arusha Declaration Ten Years After." Dar es Salaam: Government Printer, 1977. https://www.juliusnyerere.org/resources/view/the_arusha_declaration_ten_years_after_julius_k._nyerere_1977.

Nyerere, Julius. *Decentralization.* Dar es Salaam: Government Printer, 1972.

Nyerere, Julius. *Five Years of CCM Government: The Address Given to the National Conference of Chama Cha Mapinduzi by the Chairman, Ndugu Julius K. Nyerere, on October 20th 1982 at Diamond Jubilee Hall, Dar es Salaam.* Dar es Salaam: Government Printer, 1982.

Nyerere, Julius K. *Freedom and Development/Uhuru Na Maendeleo: A Selection from Writings and Speeches 1968–1973.* Oxford: Oxford University Press, 1973.

Nyerere, Julius. *President Nyerere's Speech in Parliament, 18th July, 1975.* Dar es Salaam: Government Printer, 1975. https://www.juliusnyerere.org/resources/view/president_nyerere_speech_to_parliament_18th_july_1975.

Nyerere, Julius. *Ujamaa: Essays on Socialism.* Oxford: Oxford University Press, 1968.

Nyerere, Julius. *Ujamaa Vijijini.* Dar es Salaam: Jamhuri ya Muungano, 1967.

Nyerere, Julius. "Unity for a New Order." Speech delivered February 12, 1979, Arusha, Tanzania. http://www.juliusnyerere.org/uploads/unity_for_a_new_order
   _1979.pdf.

O'Barr, Jean F. "Cell Leaders in Tanzania." *African Studies Review* 15, no 3 (1972):
   437–65.

O'Keefe, Phil, and John Soussan. "Energy: Power to Some People." *Review of African
   Political Economy* 51 (1991): 107–14.

Openshaw, Keith, and Food and Agriculture Organization of the United Nations.
   *Forest Industries Development Planning: Tanzania: Present Consumption and
   Future Requirements of Wood in Tanzania.* Rome: United Nations Development
   Programme, 1971.

Owens, Geoffrey Ross. "From Collective Villages to Private Ownership: Ujamaa,
   Tamaa, and the Postsocialist Transformation of Peri-urban Dar es Salaam,
   1970–1990." *Journal of Anthropological Research* 70, no. 2 (2014): 207–31.

Owens, Geoffrey Ross. "Post-colonial Migration: Virtual Culture, Urban Farming
   and New Peri-growth in Dar es Salaam, Tanzania, 1975–2000." *Africa: Journal of
   the International African Institute* 80, no. 2 (2010): 249–74.

PADCO. *A Proposal for an Urban Development Corporation in Tanzania.* Washington,
   DC: PADCO, 1969.

Paddison, Ronan. "Ideology and Urban Primacy in Tanzania." CURR Discussion
   Paper No. 35, University of Glasgow Centre for Urban and Regional Research,
   June 1988.

Parnell, Susan, and Edgar Pieterse, eds. *Africa's Urban Revolution.* New York: Zed
   Books, 2014.

Perkins, F. C. "Technology Choice, Industrialisation and Development Experiences
   in Tanzania." *Journal of Development Studies* 19, no. 2 (1983): 213–43.

Persson, Reidar. *Forest Resources of Africa: An Approach to International Forest Resource Appraisals.* Stockholm: Royal College of Forestry, 1977.

Perullo, Alex. *Live from Dar es Salaam: Popular Music and Tanzania's Music Economy.*
   Bloomington: Indiana University Press, 2011.

Peter, Christian, and Klas Sander. *Environmental Crisis or Sustainable Development
   Opportunity? Transforming the Charcoal Sector in Tanzania: A Policy Note.* Washington, DC: World Bank, 2009. http://documents.worldbank.org/curated/
   en/610491468122077612/pdf/502070WP0Polic1BOx0342042B01PUBLIC1.
   pdf.

Peterson, Derek. *Ethnic Patriotism and the East African Revival: A History of Dissent, c.
   1935–1972.* Cambridge: Cambridge University Press, 2012.

Peterson, Derek, Emma Hunter, and Stephanie Newell. *African Print Cultures: Newspapers and Their Publics in the Twentieth Century.* Ann Arbor: Michigan University Press, 2016.

Ponte, Stefano. *Farmers and Markets in Tanzania: How Policy Reforms Affect Rural
   Livelihoods in Africa.* Portsmouth, NH: Heinemann, 2002.

Porter, Andrew N. *The Imperial Horizons of British Protestant Missions, 1880–1914.* Grand Rapids, MI: William B. Eerdmans, 2003.

Porter, Gina. "Living in a Walking World: Rural Mobility and Social Equity Issues in Sub-Saharan Africa." *World Development* 30, no. 2 (2002): 285–300.

Potts, Deborah. "Shall We Go Home? Increasing Urban Poverty in African Cities and Migration Processes." *Geographical Journal* 161, no. 3 (1995): 245–64.

Powdermaker, Hortense. *Copper Town: Changing Africa: The Human Situation on the Rhodesian Copperbelt.* New York: Harper and Row, 1962.

Project Planning Associates. *National Capital Master Plan, Dar es Salaam, United Republic of Tanzania.* Toronto: Project Planning Associates, 1968.

Project Planning Associates. *National Capital Master Plan, Dodoma, Tanzania.* Toronto: Project Planning Associates Limited, 1976.

Project Planning Associates. "Recommended Capital Works Programme." Supplement, *National Capital Master Plan Dar es Salaam, United Republic of Tanzania.* Toronto: Project Planning Associates, 1968.

Project Report for "Waste Disposal in Dar es Salaam." In the *2nd year East African Society and Environment Papers.* University of Dar es Salaam: Faculty of Arts and Social Sciences, 1975–1976.

Pursell, Caroll. "Appropriate Technology, Modernity, and U.S. Foreign Aid." In *Science and Cultural Diversity: Proceedings of the XXIst International Congress of Science,* 175–87. Mexico City: Mexican Society for the History of Science and Technology and the National Autonomous University of Mexico, 2003.

Quayson, Ato. *Oxford Street, Accra: City Life and the Itineraries of Transnationalism.* Durham, NC: Duke University Press, 2014.

Rademacher, Anne. "Urban Political Ecology." *Annual Review of Anthropology* 44 (2015): 137–52.

Ralph, Michael. "Killing Time." *Social Text* 26, no. 4 (2008): 1–29.

Regional and Country Studies Branch Division for Industrial Studies. *United Republic of Tanzania.* Industrial Development Review Series UNIDO/IS.628, United Nations Industrial Development Organization, April 29, 1986. https://open.unido.org/api/documents/4810235/download/UNITED%20REPUBLIC%20OF%20TANZANIA.%20INDUSTRIAL%20DEVELOPMENT%20REVIEW%20SERIES%20%2815595.en%29.

Reid, Donald. *Paris Sewers and Sewermen: Realities and Representations.* Cambridge, MA: Harvard University Press, 1993.

Rizzo, Matteo. "Being Taken for a Ride: Privatisation of the Dar Es Salaam Transport System 1983–1998." *Journal of Modern African Studies* 40 (2002): 133–57.

Rizzo, Matteo. "'Life Is War': Informal Transport Workers and Neoliberalism in Tanzania 1998–2009." *Development and Change* 42, no. 5 (2011): 1179–206.

Rizzo, Matteo. *Taken for a Ride: Grounding Neoliberalism, Precarious Labour, and Public Transport in an African Metropolis.* Oxford: Oxford University Press, 2017.

Roberts, George. "Politics, Decolonization and the Cold War in Dar es Salaam 1965–1972." PhD diss., University of Warwick, 2016.

Robertson, Claire. *Trouble Showed the Way: Women, Men, and Trade in the Nairobi Area, 1890–1990*. Bloomington: Indiana University Press, 1997.

Robertson, Thomas. *Malthusian Moment: Global Population Growth and the Birth of American Environmentalism*. New Jersey: Rutgers University Press, 2012.

Robertson, Thomas. "'This Is the American Earth': American Empire, the Cold War, and American Environmentalism." *Diplomatic History* 32, no. 4 (2008): 561–84.

Robinson, Jennifer. "Global and World Cities: A View from off the Map." *International Journal of Urban and Regional Research* 26, no. 3 (2002): 531–54.

Roitman, Janet. *Anti-crisis*. Durham, NC: Duke University Press, 2013.

Rostow, W. W. *The Stages of Economic Growth: A Non-Communist Manifesto*. Cambridge: Cambridge University Press, 1960.

Roy, Ananya, and Aihwa Ong, eds. *Worlding Cities: Asian Experiments and the Art of Being Global*. Malden, MA: Wiley-Blackwell, 2011.

Sabin, Paul. *The Bet: Paul Ehrlich, Julian Simon, and Our Gamble over Earth's Future*. New Haven, CT: Yale University Press, 2013.

Sabot, R. H. *Economic Development and Urban Migration: Tanzania, 1900–1971*. Oxford: Clarendon, 1978.

Sachs, Wolfgang. *The Development Dictionary: A Guide to Knowledge as Power*. New York: Zed Books, 2010.

Samoff, Joel. "The Bureaucracy and the Bourgeoisie: Decentralization and Class Structure in Tanzania." *Comparative Studies in Society and History* 21, no. 1 (1979): 30–62.

Samoff, Joel. "Crises and Socialism in Tanzania." *Journal of Modern African Studies* 19, no. 2 (1981): 279–306.

Sassen, Saskia. "Cityness in the Urban Age." *Urban Age* Bulletin 2, Autumn 2005, 1–3.

Saul, John S. *A Flawed Freedom: Rethinking Southern African Liberation*. Chicago: University of Chicago Press, 2014.

Saul, John S. *Revolutionary Traveller: Freeze-Frames from a Life*. Winnipeg: Arbeiter Ring, 2009.

Savage, Michael, Gaynor Bagnall, and Brian J. Longhurst. *Globalization and Belonging*. London: Sage, 2005.

Sawio, Camillus J. *Urban Agriculture and the Sustainable Dar es Salaam Project*. Ottawa: International Development Research Centre, 1994.

Schell, Eileen E. "Transnational Environmental Justice Rhetorics and the Green Belt Movement: Wangari Muta Maathai's Ecological Rhetorics and Literacies." *JAC* 33, no. 3/4 (2013): 585–613.

Schmetzer, Hartmut. "Housing in Dar-es-Salaam." *Habitat International* 6, no. 4 (1982): 497–511.

Schneider, Leander. *Government of Development: Peasants and Politicians in Postcolonial Tanzania*. Bloomington: Indiana University Press, 2014.

Schneider, Leander. "The Maasai's New Clothes: A Developmentalist Modernity and Its Exclusions." *Africa Today* 53, no. 1 (2006): 101–31.

Scholz, Wolfgang, Peter Robinson, and Tanya Dayaram. "Colonial Planning Concept and Post-colonial Realities: The Influence of British Planning Culture in Tanzania, South Africa and Ghana." In *Urban Planning in Sub-Saharan Africa: Colonial and Post-Colonial Planning Cultures*, edited by Carlos Nunes Silva, 67–94. New York: Routledge, 2015.

Schumacher, E. F. *Small Is Beautiful: Economics as if People Mattered*. London: Blond and Briggs, 1973.

Schwenkel, Christina. "Post/Socialist Affect: Ruination and Reconstruction of the Nation in Urban Vietnam." *Cultural Anthropology* 28, no. 2 (2013): 252–77.

Scott, James C. *Against the Grain: A Deep History of the Earliest States*. New Haven, CT: Yale University Press, 2017.

Scott, James C. *Seeing Like a State: How Certain Schemes to Improve the Human Condition Have Failed*. New Haven, CT: Yale University Press, 1999.

Segal, Edwin S. "Ethnic Variables in East African Urban Migration." *Urban Anthropology* 2, no. 2 (1973): 194–204.

Serlin, David. "Confronting African Histories of Technology: A Conversation with Keith Breckenridge and Gabrielle Hecht." *Radical History Review* 2017, 127 (2017): 87–102.

Shechambo, Fanuel C. "Urban Demand for Charcoal in Tanzania: Some Evidence from Dar es Salaam and Mwanza." Research Report No. 67, Institute of Resource Assessment, University of Dar es Salaam, May 1986. Herskovits Library, Northwestern University.

Shipton, Parker. "African Famines and Food Security: Anthropological Perspectives." *Annual Review of Anthropology* 19, no. 1 (1990): 353–94.

Shipton, Parker. *Mortgaging the Ancestors: Ideologies of Attachment in Africa*. New Haven, CT: Yale University Press, 2009.

Shivji, Issa G. *Class Struggles in Tanzania*. London: Heinemann, 1976.

Shivji, Issa G., et al. *Report of the Presidential Commission of Inquiry into Land Matters*. Vol. 1, *Land Policy and Land Tenure Structure*. Dar es Salaam: Ministry of Lands, Housing and Urban Development, in cooperation with Scandinavian Institute of African Studies, 1994.

Siegelbaum, Lewis H. *The Socialist Car: Automobility in the Eastern Bloc*. Ithaca, NY: Cornell University Press, 2011.

Simone, AbdouMaliq. *City Life from Jakarta to Dakar: Movements at the Crossroads*. New York: Routledge, 2010.

Simone, AbdouMaliq. *For the City Yet to Come: Changing African Life in Four Cities*. Durham, NC: Duke University Press, 2004.

Simone, AbdouMaliq. "On the Worlding of African Cities." *African Studies Review* 44, no. 2 (2001): 15–41.

Simone, AbdouMaliq. "People as Infrastructure: Intersecting Fragments in Johannesburg." *Public Culture* 16, no. 3 (2004): 407–29.

Simone, AbdouMaliq. "Pirate Towns: Reworking Social and Symbolic Infrastructures in Johannesburg and Douala." *Urban Studies* 43, no. 2 (2006): 357–70.

Simone, AbdouMaliq. "Waiting in African Cities." In *Indefensible Space: The Architecture of the National Insecurity State*, edited by Michael Sorkin, 97–109. New York: Routledge, 2007.

Simone, AbdouMaliq, and Edgar Pieterse. *New Urban Worlds: Inhabiting Dissonant Times*. Medford, MA: Polity, 2017.

Singer, H. W. "Rural Unemployment as a Background to Rural-Urban Migration in Africa." *Manpower and Unemployment Research in Africa* 6, no. 2 (1973): 37–45.

Sir Alexander Gibb and Partners. *A Plan for Dar es Salaam: Report*. London: Sir Alexander Gibb and Partners, 1949.

Skutsch, Margaret McCall. "Why People Don't Plant Trees: The Socioeconomic Impacts of Existing Woodfuel Programs: Village Case Studies, Tanzania." Discussion Paper D-73P in Energy in Developing Countries Series, Resources for the Future, Washington, DC, 1983.

Smith, William Edgett. "President I: We Can't Go to the Moon!" *New Yorker*, October 16, 1971, 42–43.

Smith, William Edgett. "President III: We Must Run While They Walk." *New Yorker*, October 30, 1971, 53.

"So, Firewood Can Wreck a Home?" Kenya Woodfuel Development Programme, Ministry of Energy and Regional Development, 1985.

Southall, Aidan William, and Peter Claus Wolfgang Gutkind. *Townsmen in the Making: Kampala and its Suburbs*. Kampala: East African Institute of Social Research, 1957.

Spector, Julian. "Why Reston, Virginia, Still Inspires Planners 50 Years Later." *CityLab*, March 23, 2016. http://www.citylab.com/design/2016/03/reston-virginia-urban-planning-suburbs-robert-simon/474729/.

Springer, Zvonko. "Wazo Hill Cement Works Construction: 1964–1967: 'Twiga' Tanzania Cement Co., Ltd. Dar-es-Salaam." http://www.cosy.sbg.ac.at/~zzs pri/stories_from_professional_life/Wazo_Hill_Works.htm.

Star, Susan Leigh, and James Griesemer. "Institutional Ecology, 'Translations' and Boundary Objects: Amateurs and Professionals in Berkeley's Museum of Vertebrate Zoology, 1907–39." *Social Studies of Science* 19, no. 3 (1989): 387–420.

Steinbach, Daniel Rouven. "Carved Out of Nature: Identity and Environment in German Colonial Africa." In *Cultivating the Colonies: Colonial States and Their Environmental Legacies*, edited by Christina Folke Ax, Niels Brimnes, Niklas Thode Jensen, and Karen Oslund, 47–77. Athens: Ohio University Press, 2014.

Stoler, Ann Laura. *Along the Archival Grain: Epistemic Anxieties and Colonial Common Sense* Princeton, NJ: Princeton University Press, 2010.

Stoler, Ann Laura, ed. *Imperial Debris: On Ruins and Ruination*. Durham, NC: Duke University Press, 2013.

Strasser, Susan. *Waste and Want: A Social History of Trash*. New York: Metropolitan Books, 1999.

Stren, Richard. "The Administration of Urban Services." In *African Cities in Crisis: Managing Rapid Urban Growth*, edited by Richard E. Stren and Rodney R. White, 37–67. Boulder, CO: Westview, 1989.

Stren, Richard E. "Ujamaa Vijijini and Bureaucracy in Tanzania." *Canadian Journal of African Studies* 15, no. 3 (1981): 591–98.

Stren, Richard. "Underdevelopment, Urban Squatting and State Bureaucracy: A Case Study of Tanzania." *Canadian Journal of African Studies* 16, no. 1 (1982): 67–91.

Stren, Richard, Mohamed Halfani, and Joyce Malombe. "Coping with Urbanization and Urban Policy." In *Beyond Capitalism vs. Socialism in Kenya and Tanzania*, edited by Joel D. Barkan, 175–200. Boulder, CO: Lynne Rienner, 1994.

Stren, Richard E., and Rodney R. White. *African Cities in Crisis: Managing Rapid Urban Growth*. Boulder, CO: Westview, 1989.

Sturgis, Sam. "The Bright Future of Dar es Salaam, an Unlikely African Megacity." *CityLab*, February 25, 2015. Accessed August 16, 2019. https://www.citylab.com/design/2015/02/the-bright-future-of-dar-es-salaam-an-unlikely-african-megacity/385801/.

Sturmer, Martin. *The Media History of Tanzania*. Mtwara: Ndanda Mission, 1998.

Sugrue, Thomas. *The Origins of the Urban Crisis: Race and Inequality in Postwar Detroit*. Princeton, NJ: Princeton University Press, 1996.

Sunseri, Thaddeus. "'Every African a Nationalist': Scientific Forestry and Forest Nationalism in Colonial Tanganyika." *Comparative Studies in Society and History* 49, no. 4 (2007): 883–913.

Sunseri, Thaddeus. "'Something Else to Burn': Forest Squatters, Conservationists, and the State in Modern Tanzania." *Journal of Modern African Studies* 43, no. 4 (December 2005): 609–40.

Sunseri, Thaddeus. *Wielding the Ax: State Forestry and Social Conflict in Tanzania, 1820–2000*. Athens: Ohio University Press, 2009.

Sutton, J. E. G. "Dar es Salaam: A Sketch of a Hundred Years." *Tanzania Notes and Records* 71 (1970): 1–18.

Swanson, Maynard W. "The Sanitation Syndrome: Bubonic Plague and Urban Native Policy in the Cape Colony, 1900–1909." *Journal of African History* 18, no. 3 (1977): 387–410.

Swantz, Marja-Liisa, and Deborah Fahy Bryceson. *Women Workers in Dar Es Salaam: 1973/74 Survey of Female Minimum Wage Earners and Self-Employed*. Research Paper no. 43, Bureau of Resource Assessment and Land Use Planning, University of Dar es Salaam, 1976.

TANU. *Mwongozo wa TANU*. Dar es Salaam: Government Printer, 1971.

Tarimo, E. D. M. *The Effects of Trade Liberalisation on Property Development: Case Study [of] Dar es Salaam*. Dar es Salaam: Ardhi Institute, 1989.

Tarr, Joel. *The Search for the Ultimate Sink: Urban Pollution in Historical Perspective*. Akron: University of Akron Press, 1996.

Taylor, Peter J. "How Do We Know We Have Global Environmental Problems? Undifferentiated Science-Politics and Its Potential Reconstruction." *Changing Life: Genomes, Ecologies, Bodies, Commodities*, edited by Peter J. Taylor, Saul E. Halfon, and Paul N. Edwards, 149–74. Minneapolis: University of Minnesota Press, 1997.

Temple, Paul H. "Aspects of the Geomorphology of the Dar es Salaam Area." *Tanzania Notes and Records* 71 (1970): 21–54.

Temu, A. B., B. K. Kaale, J. A. Maghembe, ed. "Wood-Based Energy for Development in Tanzania." Proceedings of a national seminar sponsored by The Ministry of Natural Resources and Tourism with support from The Swedish International Development Agency (SIDA), Dar es Salaam, Tanzania, March 26–28, 1984.

Thomas, Lynn M. "The Modern Girl and Racial Respectability in 1930s South Africa." *Journal of African History* 47, no. 3 (2006): 461–90.

Thomas, Lynn M. "Modernity's Failings, Political Claims, and Intermediate Concepts." *American Historical Review* 116, no. 3 (2011): 727–40.

Tostensen, Arne, Inge Tvedten, and Mariken Vaa, eds. *Associational Life in African Cities: Popular Responses to the Urban Crisis*. Uppsala: Nordiska Afrikainstitutet, 2001.

Tripp, Aili Mari. *Changing the Rules: The Politics of Liberalization and the Urban Informal Economy in Tanzania*. Berkeley: University of California Press, 1997.

Tripp, Aili Mari. "Defending the Right to Subsist: The State vs. the Urban Informal Economy in Tanzania." Working Paper no. 59, World Institute for Development Economics Research of the United Nations University, August 1989.

Tripp, Aili Mari. "Urban Farming and Changing Rural-Urban Interactions in Tanzania." In *What Went Right in Tanzania: People's Response to Directed Development*, edited by Marja Liisa Swantz and Aili Mari Tripp, 98–116. Dar es Salaam: University of Dar es Salaam Press, 1996.

Tsing, Anna Lawenhaupt. "On Nonscalability: The Living World is Not Amenable to Precision-Nested Scales." *Common Knowledge* 18, no. 3 (2012): 505–24.

Twine, Wayne C., and Ricardo M. Holdo. "Fuelwood Sustainability Revisited: Integrating Size Structure and Resprouting into a Spatially Realistic Fuelshed Model." *Journal of Applied Ecology* 53, no. 6 (2016): 1766–76.

United Nations. *Report of the United Nations Conference on New and Renewable Sources of Energy, Nairobi, 10 to 21 August 1981*. A/CONF.100/11, United Nations, May 1982. https://digitallibrary.un.org/record/25034/files/A_CONF-100_11 -EN.pdf.

United Nations. *Resolution Adopted by the General Assembly: 44/172. Plan of Action to Combat Desertification*. United Nations A/RES/44/172, December 19, 1989. http://www.un-documents.net/a44r172.htm.

Vale, Lawrence J. *Architecture, Power, and National Identity*. New Haven, CT: Yale University Press, 1992.

van Ginneken, Sophie. "The Burden of Being Planned. How African Cities Can Learn from Experiments of the Past: New Town Dodoma, Tanzania." International New Town Institute. Accessed September 9, 2017. http://www.newtown institute.org/spip.php?article1050.

Vassanji, M. G. *And Home Was Kariakoo: A Memoir of East Africa*. Toronto: Doubleday Canada, 2014.

Vazifdar, J. S. "The Cement Industry in Tanzania." *Tanzania Notes and Records* 81 and 82 (1977): 125–34.

Veal, Michael. *Fela: Life and Times of an African Musical Icon*. Philadelphia: Temple University Press, 2000.

Vestbro, Dick Urban. *Social Life and Dwelling Space: An Analysis of Three House Types in Dar Es Salaam*. Lund: Institutionen för Byggnadsfunktionslära, Tekniska Högskolan i Lund, 1975.

von Schnitzler, Antina. *Democracy's Infrastructure: Techno-Politics and Protest after Apartheid*. Princeton, NJ: Princeton University Press, 2016.

Wainaina, Binyavanga. "Glory." *Bidoun Magazine*. Issue #10, Technology. http://archive.bidoun.org/magazine/10-technology/.

Wakeman, Rosemary. *Practicing Utopia: An Intellectual history of the New Town Movement*. Chicago: University of Chicago Press, 2016.

Wampole, Christy. *Rootedness: The Ramifications of a Metaphor*. Chicago: University of Chicago Press, 2016.

Webb, Lawrence. *Cinema of Urban Crisis: Seventies Film and the Reinvention of the City*. Amsterdam: Amsterdam University Press, 2014.

Weinstein, Liza. *Durable Slum: Dharavi and the Right to Stay Put in Globalizing Mumbai*. Minneapolis: University of Minnesota Press, 2014.

Wells, Jill. "The Construction Industry in the Context of Development: A New Perspective." *Habitat International* 8, no. 3 (1984): 9–28.

White, Luise. *Comforts of Home: Prostitution in Colonial Nairobi*. Chicago: University of Chicago Press, 1990.

White, Luise. "Hodgepodge Historiography: Documents, Itineraries, and the Absence of Archives." *History in Africa* 42 (2015): 309–18.

White, Luise. *Speaking with Vampires: Rumor and History in Colonial Africa*. Berkeley: University of California Press, 2000.

White, Richard. *Railroaded. The Transcontinentals and the Making of Modern America*. New York: W. W. Norton, 2012.

Wield, David, and Carol Barker. "Course Bibliography. Science, Technology and Development: Part of a Course in Development Studies for First and Second Year

Engineering and Medical Students at the University of Dar Es Salaam.'" *Social Studies of Science* 8, no. 3 (1978): 385–95.

Williams, Raymond. *The Country and the City*. New York: Oxford University Press, 1975.

Williams, Raymond. *Keywords: A Vocabulary of Culture and Society*. New York: Oxford University Press, 1985.

The Wives of Nendi. Directed by Stephen Peet. Central African Film Unit, 1949. http://www.colonialfilm.org.uk/node/312.

World Bank. *Report and Recommendation of the President of the International Bank for Reconstitution and Development to the Executive Directors on a Proposed Loan to the United Republic of Tanzania for an Urban Water Supply Project*. Report no. P-1928a-TA, World Bank, December 7, 1976. http://documents.worldbank.org/curated/en/469301468116372821/pdf/multi0page.pdf.

World Bank. "Staff Appraisal Report." Dar es Salaam Sewerage and Sanitation Project. Report no. 3678-TA, World Bank, November 29, 1982.

World Bank. *Tanzania Appraisal of National Sites and Services Project*. Report no. 337a-TA, World Bank, May 12, 1974.

World Bank. *Tanzania: Issues and Options in the Energy Sector*. Report no. 4969-TA, World Bank, November 1984. http://documents.worldbank.org/curated/en/403251468778743225/Tanzania-Issues-and-options-in-the-energy-sector.

World Bank. *Tanzania: The Second National Sites and Services Project*. Annex 1. Report no. 1518a-TA, World Bank, June 9, 1977.

Yeager, Rodger. "Demography and Development Policy in Tanzania." *Journal of Developing Areas* 16, no. 4 (1982): 489–510.

Yhdego, Michael. "Industrialization and Environmental Pollution in Tanzania. Case Studies: Wazo Hill Cement Factory and Tanzania Fertilizer Company." PhD diss., Aalbord University of Denmark, 1988.

Yhdego, Michael. "Scavenging Solid Wastes in Dar es Salaam, Tanzania." *Waste Management Research* 9, no. 4 (1991): 259–65.

Zollner, Douglas. "A Village Woodlot Project in Tanzania." *Arid Lands Newsletter* 13 (1981).

INDEX

Note: Page numbers in *italics* refer to illustrative materials.

94, 95; materials of, 64; saving,
121; Tanzanian, 104–5, 183; urban,
100
Sokoine, Edward, 141, 142, 143–44,
208n76
sovereignty, 176; economic, 12, 137;
foreign, 176; materials of, 64;
national, 84; political, 12;
resource, 131–32, 151, 173, 179
space, 136; claiming, 37; development
across, 77; interstitial, 146; living,
66; open, 26, 27, 98, 118, 195n28,
212n36; peri-urban, 58; production, 147; public, 94, 103, 104;
reclamation of, 99; transgressive,
118; unplanned, 61. *See also* urban
space
spare parts, 31, 75, 92, 99, 101, 109, 110,
111, 113, 128, 131, 137, 139, 140,
150, 209n89
squatter areas, 44, 48, 153
stewardship, 136, 173; hoarding and,
148
Stoler, Ann Laura, 69, 193n1
stoves, 166, 167, 180
Strasser, Susan, 132, 215n78
subsistence practices, 93, 122, 126, 153,
154, 155, 160, 162, 163, 170
Sunseri, Thaddeus, 175, 220n10
Survey of Dar es Salaam, A (Leslie), 38
sustainability, 149, 151, 176, 178
Swahili, 3, 67
Swahili coast, 66, 151

Tabata Development Fund, 128,
213n58
TANESCO, 125, 136
Tanga, 47, 77
Tanganyika, 16, 152; Zanzibar and, 16
Tanganyika Packers, 46, 58
Tanganyikan African National
(TANU), 21, 23, 31, 34, 46, 53,

54, 62, 78, 117, 120, 123, 135, 140,
142, 143, 153; basic organ of, 49;
campaign by, 144–45; capital relocation and, 19, 22; guidelines for,
104–8; language of, 96, 105; SIDO
and, 133; walanguzi and, 144
Tanzania: capital of, 20–23; emergence
of, 16; map of, *17*
Tanzanian Housing Bank (THB), 45,
47
TAZARA railway, 71, 209n96
technology, 6, 64, 65, 80, 82, 83, 96,
111, 161, 162, 178; appropriate, 12,
46, 78, 136, 139, 165–70, 173; black
box, 139; cement, 77; decisions
about, 138–139; dilemmas of,
138, 146, 165; foreign, 177, 178;
imported, 95, 139; Northern, 177;
rethinking, 136–40; science and,
9, 10, 162, 164; transfer of, 81, 138;
transformative, 173; Western, 138,
165–66
technopolitics, 65–71, 199n8
Temeke, 30, 39, 40, 52, 129, 194n16
ten-cell unit, 48–61; sketch of, *48*
THB. *See* Tanzanian Housing Bank
Third World, 6, 12, 16, 81, 82, 95, 121,
137, 150, 151, 161, 162, 164, 181,
188n12, 221n57; development and,
177; solidarities of, 176
transportation, 8, 13, 18, 61, 65, 76,
125, 184; breakdown of, 109, 112,
134; public, 6, 14, 91–92, 97, 99,
100, 101, 102, 103–4, 107, 124,
206n47, 206n48
trees, 156, 169; felling of, *168*; planting,
165, 170–173, *171*, 224n121
"Trouble at Hill" (Ayres), 62

Ubungo, 97, 99, 102, 117, 165
UDA. *See* Usafiri Dar es Salaam
UDSM. *See* University of Dar es Salaam